Advance Praise for *Catching Fire*

'In modern times we are all obsessed with how to cook, but this book answers the much deeper question of *why* we cook, and in doing so highlights the fact that what has become an art form, was at its inception the driving force for us, humans-to-be, becoming the dominant species on the planet.

Wrangham's explanations are always thorough without feeling like you're swimming through treacle. They are simple, logical and compelling, whilst answering some of the biggest questions out there.

At a time when less people are cooking, this book gently reminds us that we in the developed world are walking away from the very thing that made us what we are, and we should squander this defining skill at our own peril.' Allegra McEvedy, *Guardian*

'A feast of new ideas on human evolution' Steven Pinker, author of *How the Mind Works* and *The Stuff of Thought*

'Richard Wrangham is the perfect master of paleoanthropology, primatology, archaeology, human biology, and the chemistry and physics of food. No one else could combine these disciplines to yield revolutionary insights about food history' Felipe Fernandez-Armesto, author of *The World: A Global History*

'A *tour de force* on how to study human evolution, combining original ideas with an extraordinary range of science. With elegance and clarity, Wrangham has shown how cooking permeates all human life, and must have played a major part in making us what we are as a species' Robert Foley, Director of the Centre for Human Evolutionary Studies, University of Cambridge

'Superbly lucid and comprehensive ... this masterful work shows how cooking was – and continues to be – an essential part of humanity' David Pilbeam, Professor of Human Evolution, Harvard University

'The claim that "cooks made us" now has exciting evolutionary support. Let cooks lead the celebrations!' Michael Symons, author of *A History of Cooks and Cooking*

'*Catching Fire* is a brilliant, pathbreaking book. Every reader will be inspired by it' Nicholas Humphrey, author *of The Mind Made Flesh and Seeing Red*

'This is a fascinating tour through an everyday event so important we hardly notice, cooking: a human habit for some two million years, which turns out to be a key to much of who we are. Beautifully written, with convincing logic and evidence throughout' Robert Trivers, winner of the 2007 Crafoord Prize

'*Catching Fire* is convincing in argument and impressive in its explanatory power. A rich and important book' Michael Pollan, author of *In Defense of Food* and *The Omnivore's Dilemma*

Catching Fire

RICHARD WRANGHAM is the Ruth Moore Professor of Biological Anthropology at Harvard University, Curator of Primate Behavioral Biology at the Peabody Museum, and Director of the Kibale Chimpanzee Project in Uganda. He is the co-author of *Demonic Males* and co-editor of *Primate Societies, Chimpanzee Cultures, Science and Conservation in African Forests* and *Sexual Coercion in Primates*. He lives in Cambridge, Massachusetts.

Catching Fire

How Cooking Made Us Human

RICHARD WRANGHAM

PROFILE BOOKS

First published in Great Britain in 2009 by
PROFILE BOOKS LTD
3A Exmouth House
Pine Street
London EC1R 0JH
www.profilebooks.com

First published in the United States of America by
Basic Books, a member of the Perseus Books Group

10 9 8 7 6 5 4 3 2 1

Designed by Timm Bryson

Printed and bound in Great Britain by
Clays, Bungay, Suffolk

A CIP catalogue record for this book is available from the British Library.

ISBN 978 1 84668 285 8

The paper this book is printed on is certified by the © 1996 Forest Stewardship Council A.C. (FSC). It is ancient-forest friendly. The printer holds FSC chain of custody SGS-COC-2061

FSC
Mixed Sources
Product group from well-managed forests and other controlled sources

Cert no. SGS-COC-2061
www.fsc.org
© 1996 Forest Stewardship Council

CONTENTS

INTRODUCTION

The Cooking Hypothesis

> *"[Fire] provides us warmth on cold nights; it is the means by which they prepare their food, for they eat nothing raw save a few fruits . . . the Andamanese believe it is the possession of fire that makes human beings what they are and distinguishes them from animals."*
>
> —A. R. Radcliffe-Brown, *The Andaman Islanders: A Study in Social Anthropology*

The question is old: Where do we come from? The ancient Greeks told of human shapes being molded by gods out of clay. We know now that our bodies were molded by natural selection and that we come from Africa. In the distant past, long before people first wrote or tilled the soil or took to boats, our ancestors lived there as hunters and gatherers. Fossilized bones reveal our kinship with ancient Africans a million years ago and more, people who looked much like we do today. But in deeper rocks the record of our humanity dwindles until around two million years ago, when it gives way to prehuman ancestors and leaves us with a question that every culture answers in a different way, but only science can truly decide. What made us human?

This book proposes a new answer. I believe the transformative moment that gave rise to the genus *Homo,* one of the great transitions in the history of life, stemmed from the control of fire and the advent of cooked meals. Cooking increased the value of our food. It changed our bodies, our brains, our use of time, and our social lives. It made us into consumers of external energy and thereby created an organism with a new relationship to nature, dependent on fuel.

The fossil record shows that before our ancestors came to look like us, they were humanlike in walking upright, but mostly they had the characteristics of nonhuman apes. We call them australopithecines. Australopithecines were the size of chimpanzees, they climbed well, they had ape-sized bellies, and they had protruding, apelike muzzles. Their brains, too, were barely larger than those of chimpanzees, which suggests that they would have been as uninterested in the reasons for their existence as the antelopes and predators with which they shared their woodlands. If they still lived today in some remote area of Africa, we would find them fascinating. But to judge from their ape-sized brains, we would observe them in national parks and keep them in zoos, rather than give them legal rights or invite them to dinner.

Although the australopithecines were far different from us, in the big scheme of things they lived not so long ago. Imagine going to a sporting event with sixty thousand seats around the stadium. You arrive early with your grandmother, and the two of you take the first seats. Next to your grandmother sits her grandmother, your great-great-grandmother. Next to her is your great-great-great-great-grandmother. The stadium fills with the ghosts of preceding grandmothers. An hour later the seat next to you is occupied by the last to sit down, the ancestor of you all. She nudges your elbow, and you turn to find a strange nonhuman face. Beneath a low forehead and big brow-ridge, bright dark eyes surmount a massive jaw. Her long, muscular arms and short legs intimate her gymnastic climbing ability. She is your ancestor and an australopithecine, hardly a companion your grandmother can be expected to enjoy. She grabs an overhead beam and swings away over the crowd to steal some peanuts from a vendor.

She is connected to you by over three million years of rain and sun and searching for food in the rich and scary African bush. Most australopithecines eventually went extinct but her lineage slowly changed. Evolutionarily, she was one of the lucky ones.

The transition is first signaled at 2.6 million years ago by sharp flakes dug from Ethiopian rock. The fragments testify to cobblestones being deliberately clashed to produce a tool. Cut marks on fossil bones show that the simple knives were used to cut tongues out of dead antelope and to get hunks of meat by slicing through tendons on animal limbs. This new behavior was remarkably effective—it would have allowed them to skin an elephant quickly—and was far more skillful than anything chimpanzees do when eating meat. Knife-making suggests planning, patience, cooperation, and organized behavior.

Old bones continue the story. By around 2.3 million years ago, the first tentative record emerges of a new species, a habiline. Habilines, still poorly understood, are the "missing link" between apes and humans. They were truly missing until 1960, when Jonathan Leakey, the twenty-year-old son of paleontologist Louis Leakey and archaeologist Mary Leakey, discovered them in the form of a jaw, skull, and hand in Tanzania's Olduvai Gorge. Even now there are only six skulls that tell us the brain size of the principal species, and just two reasonably complete specimens showing their limbs, so our portraits of these intermediate beings are fuzzy. Habilines appear to have been about the same small size as australopithecines and had long arms and jutting faces, leading some people to call them apes. Yet they are thought to be the knife makers, and they had brains twice as big as those of living nonhuman apes, so others place them in the genus *Homo* and thereby call them human. In short, they show a mixture of prehuman and human characteristics. They were like upright chimpanzees with big brains, and we might guess they were just as hairy and almost as good at climbing trees.

After the habilines emerged, it took hundreds of thousands of years for the evolutionary gears to start turning rapidly again, but between 1.9 million and 1.8 million years ago, the second critical step was taken: some habilines evolved into *Homo erectus,* and with their arrival the world faced a new future.

The mental abilities of *Homo erectus* are open to question. We do not know whether they used a primitive kind of language, or how well they controlled their tempers. But *Homo erectus* looked much more like us than any prior species. They are considered to have walked and run as fluently as we do today, with the same characteristic stride that we have. Their various descendants, including Neanderthals more than a million years later, all exhibited the same form and stature. If they time-traveled to a modern city, they might suffer some sidelong glances but they could be fitted for clothes in a typical store. Their anatomy was so similar to ours that some anthropologists call them *Homo sapiens*, but most give these pioneers their own distinct name of *Homo erectus* because of such features as smaller brains and lower foreheads than are found in modern humans. Whatever we call them, their arrival marks the genesis of our physical form. They even appear to have grown and matured slowly, in the manner of modern humans. After their emergence it would be mainly a question of time and brain growth before modern humans emerged about two hundred thousand years ago.

So the question of our origins concerns the forces that sprung *Homo erectus* from their australopithecine past.

Anthropologists have an answer. According to the most popular view since the 1950s there was a single supposed impetus: the eating of meat.

Hundreds of different hunter-gatherer cultures have been described, and all obtained a substantial proportion of their diet from meat, often half their calories or more. Archaeology indicates a similar importance of meat all the way back to the butchering habilines more than two million years ago. By contrast, there is little to suggest that their predecessors, the australopithecines, were much different from chimpanzees in their predatory behavior. Chimpanzees readily grab monkeys, piglets, or small antelopes when opportunities arise, but weeks or even months can go by with no meat in their diets. Among primates we are the only dedicated carnivores, and the only ones to take meat from large carcasses.

Those smaller-brained ancestors could not have obtained meat without confronting dangerous animals. Their physical abilities often would have proved wanting. The first meat eaters certainly would have been slow, they had small bodies, their teeth and limbs made feeble weapons, and their hunting tools were probably little more than rocks and natural clubs. Greater ingenuity and improved physical prowess would have helped bring down prey. Hunters might have chased antelope on long runs until the quarry collapsed from exhaustion. Perhaps they found carcasses by watching where vultures swooped down. Predators such as saber-toothed lions brought further challenges. Teamwork might have been necessary, with some individuals in a hunting party throwing rocks to keep fearsome animals at bay while others quickly cut off hunks of meat before all retired to eat in a defensible site. So it is easy to imagine that the rise of meat eating fostered various human characteristics such as long-distance travel, big bodies, rising intelligence, and increased cooperation. For such reasons the meat-eating hypothesis, often called "Man-the-Hunter," has long been popular with anthropologists to explain the change from australopithecine to human.

But the Man-the-Hunter hypothesis is incomplete because it does not explain how hunting was possible without the economic support gathered foods provided. Among hunter-gatherers, gathering is mostly done by women and is often responsible for half the calories brought to camp. Gathering can be just as critical as hunting because men sometimes return with nothing, in which case the family must rely entirely on gathered foods. Gathering depends on abilities normally considered to be absent in australopithecines, such as carrying large bundles of food. When and why did gathering evolve? What breakthroughs in technology enabled females to gather? Or did habilines get their meat without being involved in an economy of exchange? These are core questions Man-the-Hunter leaves unanswered.

A different kind of difficulty is even more severe: the habilines show that there were two changes in the path from ape to human, not just the one implied by Man-the-Hunter. The two steps involved different kinds of transformation and occurred hundreds of thousands of years apart—one probably around 2.5 million years ago, and the second between 1.9 million and 1.8 million years ago. It makes no sense that the two kinds of change should have been prompted by the same cause.

Meat eating accounts smoothly for the first transition, jump-starting evolution toward humans by shifting chimpanzeelike australopithecines into knife-wielding, bigger-brained habilines, while still leaving them with apelike bodies capable of collecting and digesting vegetable foods as efficiently as did australopithecines. But if meat eating explains the origin of the habilines, it leaves the second transition unexplained, from habilines to *Homo erectus*. Did habilines and *Homo erectus* obtain their meat in such different ways that they evolved different kinds of anatomy? Some people think the habilines might have been primarily scavengers while *Homo erectus* were more proficient hunters. The idea is plausible, though archaeological data do not directly test it. But it does not solve a key problem concerning the anatomy of *Homo erectus*, which had small jaws and small teeth that were poorly adapted for eating the tough raw meat of game animals. These weaker mouths cannot be explained by *Homo erectus's* becoming better at hunting. Something else must have been going on.

How lucky that Earth has fire. Hot, dry plant material does this amazing thing: it burns. In a world full of rocks, animals, and living plants, dry, combustible wood gives us warmth and light, without which our species would be forced to live like other animals. It is easy to forget what life would have been like without fire. The nights would be cold, dark, and dangerous, forcing us to wait helplessly for the sun. All our food would be raw. No wonder we find comfort by a hearth.

Nowadays we need fire wherever we are. Survival manuals tell us that if we are lost in the wild, one of our first actions should be to make a fire. In addition to warmth and light, fire gives us hot food, safe water, dry clothes, protection from dangerous animals, a signal to friends, and even a sense of inner comfort. In modern society, fire might be hidden from our view, tidied away in the basement boiler, trapped in the engine block of a car, or confined in the power station that drives the electrical grid, but we still completely depend on it. A similar tie is found in every culture. To the hunting-and-gathering Andaman Islanders of India, fire is "the first thing they think of carrying when they go on a journey," "the center round which social life moves," and the possession that distinguishes humans from animals. Animals need food, water, and shelter. We humans need all those things, but we need fire too.

How long have we needed it? Few people have thought about this question. Not even Charles Darwin pursued it, though he had every reason to be interested. During his five-year voyage around the world, Darwin learned what it was like to be hungry in the wild. When camped in harsh places, such as the sodden moors of the Falkland Islands, he made fire by rubbing sticks together. He cooked with hot rocks in an earth oven and called the art of making fire "probably the greatest [discovery], excepting language, ever made by man." His gritty experiences taught him that "hard and stringy roots can be rendered digestible, and poisonous roots or herbs innocuous." He understood the value of cooked food.

But Darwin showed no interest in knowing when fire was first controlled. His passion was evolution, and he thought fire was irrelevant to how we evolved. Like most people, he simply assumed that by the time our ancestors first controlled fire they were already human. He cited his fellow evolutionist Alfred Russel Wallace approvingly: "man is enabled through his mental faculties 'to keep with an unchanged body in harmony with the changing universe.'" The control of fire was just another way for an unchanged body with an adept mental faculty to respond to a natural challenge. "When he migrates into a colder climate he uses clothes, builds sheds, and makes fires; and, by the aid of fire, cooks food otherwise indigestible . . . the lower animals, on the other hand, must have their bodily structure modified in order to survive under greatly changed conditions."

The notion of prehistoric humans having an "unchanged body" while inventing new ways to make their lives easier is mostly right. Little change has occurred in human anatomy since the time of *Homo erectus* almost two million years ago. Culture is the trump card that enables humans to adapt, and compared to the two-million-year human career, most cultural innovation has indeed been recent. Before two hundred thousand years ago, the main novelties recorded by archaeology were stone tools and spears. Art, fishing tools, personal decoration with necklaces, and stone-tipped weapons all came later. Why should the control of fire be any older? Most anthropologists have followed Darwin's assumption that cooking has been a late addition to the human skill set, a valuable tradition without any biological or evolutionary significance. We use fire, Darwin seemed to imply, but we could survive without it if we had to. The implication was that cooking has little biological importance.

A century later, cultural anthropologist Claude Lévi-Strauss produced a revolutionary analysis of human cultures that implicitly supported the biological insignificance of cooking. He was an expert on the myths of Brazilian tribes, and he was deeply impressed with the way in which cooking served to symbolize human control over nature. "Not only does cooking mark the transition from nature to culture," Lévi-Strauss wrote in his influential 1960s book, *The Raw and the Cooked*, "but through it and by means of it, the human state can be defined with all its attributes." Lévi-Strauss's insight that cooking is a defining feature of humanity was perceptive. But strikingly, for him its significance appeared to be entirely psychological. Fellow anthropologist Edmund Leach presented Lévi-Strauss's views crisply: "[People] do not have to cook their food, they do so for symbolic reasons to show that they are men and not beasts." Lévi-Strauss was an elite anthropologist, and his implication that cooking had no biological meaning was widely touted. No one challenged this aspect of his analysis.

Despite the predominant skepticism about the role of fire in human evolution, a few contrarians have argued that cooking has been a core influence on human nature. The strongest voices have come from students of food and eating. The celebrated French gastronomist Jean-Anthelme Brillat-Savarin sounded evolutionary even when Charles Darwin was still a teenager. "It is by fire that man has tamed Nature itself," he wrote in 1825. His experience told him that cooking helps us to eat meat more easily. After our ancestors started cooking, he argued, meat became more desirable and valuable, leading to a new importance for hunting. And since hunting was mainly a male activity, women took on the role of cooking. Brillat-Savarin was prescient in tracing a link from cooking to households, but his ideas were not richly developed. They were throwaway lines hidden in a voluminous output, and they have never been taken seriously.

In the past half century, ideas suggesting how the control of fire might have influenced human behavior or evolution have been proposed by writers in physical anthropology (by Carleton Coon and Loring Brace), archaeology (especially by

Catherine Perlès), and sociology (by Joop Goudsblom). But such analyses have been tentative, leaving it to the specialized field of cooking history to provide thoughts as bold as those of Brillat-Savarin. In 1998 cooking historian Michael Symons combined intellectual ingredients from a range of disciplines, and based on the idea that cooking affects many aspects of life from nutrition to society, he made a stronger claim than any before him. Symons concluded, "cooking is the missing link . . . defining the human essence. . . . I pin our humanity on cooks." In a 2001 book on the history of food, historian Felipe Fernandez-Armesto likewise declared cooking an "index of the humanity of humankind." But neither these authors nor any other writer advocating the importance of cooking understood how cooking affects the nutritional quality of food. Critical questions therefore were left untouched, such as whether humans are evolutionarily adapted to cooked food, or how cooking had its supposed effects on making us human, or when cooking evolved. The result was a series of ideas that, however intriguing, were not tied down to biological reality. They suggested that cooking had shaped us, but they did not say why or when or how.

There is a way to find out whether cooking is as biologically insignificant as Darwin implied, or as central to humanity as Symons asserts. We need to know what cooking does. Cooked food does many familiar things. It makes our food safer, creates rich and delicious tastes, and reduces spoilage. Heating can allow us to open, cut, or mash tough foods. But none of these advantages is as important as a little-appreciated aspect: cooking increases the amount of energy our bodies obtain from our food.

The extra energy gave the first cooks biological advantages. They survived and reproduced better than before. Their genes spread. Their bodies responded by biologically adapting to cooked food, shaped by natural selection to take maximum advantage of the new diet. There were changes in anatomy, physiology, ecology, life history, psychology, and society. Fossil evidence indicates that this dependence arose not just some tens of thousands of years ago, or even a few hundred thousand, but right back at the beginning of our time on Earth, at the start of human evolution, by the habiline that became *Homo erectus*. Brillat-Savarin and Symons were right to say that we have tamed nature with fire. We should indeed pin our humanity on cooks.

Those claims constitute the cooking hypothesis. They say humans are adapted to eating cooked food in the same essential way as cows are adapted to eating grass, or fleas to sucking blood, or any other animal to its signature diet. We are tied to our adapted diet of cooked food, and the results pervade our lives, from our bodies to our minds. We humans are the cooking apes, the creatures of the flame.

CHAPTER 1

Quest for Raw-Foodists

"My definition of Man is, a 'Cooking Animal'. The beasts have memory, judgement, and all the faculties and passions of our mind, in a certain degree; but no beast is a cook. . . . Man alone can dress a good dish; and every man whatever is more or less a cook, in seasoning what he himself eats."

—James Boswell, *Journal of a Tour to the Hebrides with Samuel Johnson*

Animals thrive on raw diets. Can humans do the same? Conventional wisdom has always assumed so, and the logic seems obvious. Animals live off raw food, and humans are animals, so humans should fare well on raw food. Many foods are perfectly edible raw, from apples, tomatoes, and oysters to steak tartare and various kinds of fish. Tales of raw diets are numerous. According to Marco Polo, Mongol warriors of the thirteenth century supposedly rode for ten days at a time without lighting a fire. The riders' food was the raw blood of their horses, obtained by piercing a vein. The cavalry saved time by riding without cooking, and they avoided producing the smoke that might reveal their position to hostile forces. The men did not like the liquid diet and looked forward to a cooked meal when speed was not essential, but there is no suggestion that they suffered from it. Such stories make cooking seem like a luxury, unimportant to our biological needs. But consider the Evo Diet experiment.

In 2006 nine volunteers with dangerously high blood pressure spent twelve days eating like apes in an experiment filmed by the British Broadcasting Corporation. They lived in a tented enclosure in England's Paignton Zoo and ate almost everything raw. Their diet included peppers, melons, cucumbers, tomatoes, carrots, broccoli, grapes, dates, walnuts, bananas, peaches, and so on—more than fifty kinds of fruits, vegetables, and nuts. In the second week they ate some cooked oily fish, and one man sneaked some chocolate. The regime was called the Evo Diet because it was supposed to represent the types of foods our bodies have evolved to eat. Chimpanzees or gorillas would have loved it and would have grown fat on a menu that was certainly of higher quality than they could find in the wild. The participants ate until they were full, taking in up to 5 kilograms (10 pounds) by weight per day. The daily intake was calculated by the experiment's nutritionist to include an adequate 2,000 calories for women, and 2,300 calories for men.

The aim of the volunteers was to improve their health, and they succeeded. By the end of the experiment their cholesterol levels had fallen by almost a quarter and average blood pressure was down to normal. But while medical hopes were met, an extra result had not been anticipated.

The volunteers lost a lot of weight—an average of 4.4 kg (9.7 pounds) each, or 0.37 kg (0.8 pounds) per day.

The question of what kind of diet we need is critical for understanding human adaptation. Are we just an ordinary animal that happens to enjoy the tastes and securities of cooked food without in any way depending on them? Or are we a new kind of species tied to the use of fire by our biological needs, relying on cooked food to supply enough energy to our bodies? No serious scientific tests have been designed to resolve this problem. But whereas the Evo Diet investigation was short-term and informal, a few studies of long-term raw-foodists give us systematic data with a similar result.

Raw-foodists are dedicated to eating 100 percent of their diets raw, or as close to 100 percent as they can manage. There are only three studies of their body weight, and all find that people who eat raw tend to be thin. The most extensive is the Giessen Raw Food study, conducted by nutritionist Corinna Koebnick and her colleagues in Germany, which used questionnaires to study 513 raw-foodists who ate from 70 percent to 100 percent of their diet raw. They chose to eat raw to be healthy, to prevent illness, to have a long life, or to live naturally. Raw food included not only uncooked vegetables and occasional meat, but also cold-pressed oil and honey, and some items that were lightly heated such as dried fruits, dried meat, and dried fish. Body mass index (BMI), which measures weight in relation to the square of the height, was used as a measure of fatness. As the proportion of food eaten raw rose, BMI fell. The average weight loss when shifting from a cooked to a raw diet was 26.5 pounds (12 kilograms) for women and 21.8 pounds (9.9 kilograms) for men. Among those eating a purely raw diet (31 percent), the body weights of almost a third indicated chronic energy deficiency. The scientists' conclusion was unambiguous: "a strict raw food diet cannot guarantee an adequate energy supply."

The amount of meat in the Giessen Raw Food diets was not recorded but many raw-foodists eat rather little meat. Could a low meat intake have contributed to their poor energy supply? It is possible. However, among people who eat cooked diets, there is no difference in body weight between vegetarians and meat eaters: when our food is cooked we get as many calories from a vegetarian diet as from a typical American meat-rich diet. It is only when eating raw that we suffer poor weight gain.

The energy consequences of forgoing cooked food lead to a consistent reaction, illustrated by journalist Jodi Mardesich when she became a raw-foodist. "I'm hungry. These days, I'm almost always hungry," she wrote. A typical day began at 7 A.M. when she cut and juiced two ounces of wheat grass. At 8:30 A.M. she had a bowl of "energy soup," which she describes as a "room-temperature concoction made of sunflower greens, which are the tiny first shoots of a sunflower plant, and rejuvelac, a fermented wheat drink that tastes a lot like bad lemonade." She added a couple of spoonfuls of blended papaya for interest. Lunch was a salad of sunflower greens, sprouted fenugreek seeds, sprouted broccoli seeds, fermented cabbage, and a loaf made of sprouted sunflower seeds, dehydrated seaweed, and some vegetables. Dinner was more sprouts, avocado chunks, pineapple, red onion, olive oil, raw vinegar, and sea salt. An hour later she was hungry again. In photographs she looks distinctly thin, but she was happy. She described herself as feeling energized, mentally sharper, and more serene. Nevertheless, after six months, during which she lost 18 pounds (8.2 kilograms), she could not resist slipping out for a pizza. Mardesich was not alone in finding a wholly raw diet a challenge. The Giessen Raw Food study found that 82 percent of long-term raw-foodists included some cooked food in their diets.

To judge whether the energy shortage experienced by raw-foodists is biologically significant, we need to know whether raw-induced weight loss interferes with critical functions—ideally, for a population living under conditions similar to those in our evolutionary past. In the Giessen study, the more raw food that women ate, the lower their BMI and the more likely they were to have partial or total amenorrhea. Among women eating totally raw diets, about 50 percent entirely ceased to menstruate. A further proportion, about 10 percent, suffered irregular menstrual cycles that left them unlikely to conceive. These figures are far higher than for women eating cooked food. Healthy women on cooked diets rarely fail to menstruate, whether or not they are vegetarian. But ovarian function predictably declines in women suffering from extreme energy depletion, such as marathoners and anorexics.

Raw-foodist men sometimes also report an impact on their sexual functions. In *How to Do the Raw Food Diet with Joy for Awesome Health and Success,* the author, Christopher Westra, wrote: "In my own experience, starting on living foods brought about a change in sexuality that was dramatic and completely unexpected. In just a few weeks, the number of times per day I thought about sex decreased tremendously." Westra believed that seminal emissions are designed to remove toxins from the body. After a few weeks of a raw diet, he said, the intake of toxins had fallen to the point where ejaculation was no longer necessary. In a similar way some raw-foodists regard menstruation as a mechanism for removing toxins and therefore regard its cessation as a sign of the health of their diets. Perhaps it is unnecessary to note that medical science finds no support for the idea that toxins are removed by seminal emissions or menstruation.

Reduced reproductive function means that in our evolutionary past, raw-foodism would have been much less successful than the habit of eating cooked food. A rate of infertility greater than 50 percent, such as was found in the Giessen Raw Food study, would be devastating in a natural population of foragers. And since the Giessen study was of urban people enjoying a life of middle-class ease, such dramatic effects on reproduction are mild compared to what would have happened if these German raw-foodists had been searching for food in the wild.

Most raw-foodists prepare their food elaborately in ways that increase their energy value. Techniques include mild heating, blending, grinding, and sprouting. Any system of reducing the size of food particles, such as grinding and crushing, leads to predictable increases in energy gain. The German raw-foodists also had the advantage of eating oils produced commercially by industrial processing. Koebnick's team found that about 30 percent of the subjects' calories came from these lipids, a valuable energy source that would not have been available to hunter-gatherers. Yet even with all these helpful conditions, at least half the German women eating raw foods obtained so little energy from their diet, they were physiologically unable to have babies.

The Giessen subjects had further advantages. There is no indication that they engaged in much exercise, unlike women in foraging populations. Anthropologist Elizabeth Marshall Thomas describes bushman women in Africa's Kalahari Desert returning to camp at the end of their ordinary long day thoroughly exhausted, because for much of the day they have been squatting and digging and walking, and hefting large loads of food, wood, and children. Even in populations that cook, these natural activity levels are high enough to interfere with reproductive function. If we imagine the lives of our German raw-foodists made more difficult by a daily regime of foraging for food in the wild, their rate of energy expenditure would surely be substantially increased. As a result, many more than 50 percent of the women would be incapable of pregnancy.

Then add that the subjects of the Giessen Raw Food study obtained their diets from supermarkets. Their foods were the typical products of modern farming—fruits, seeds, and vegetables all selected to be as delicious as possible. "Delicious" means high energy, because what people like are foods with low levels of indigestible fiber and high levels of soluble carbohydrates, such as sugars. Agricultural improvements have rendered fruits in a supermarket, such as apples, bananas, and strawberries, far higher in quality than their wild ancestors. In our laboratory at Harvard, nutritional biochemist NancyLou Conklin-Brittain finds that carrots contain as much sugar as the average wild fruit eaten by a chimpanzee in Kibale National Park in Uganda. But even carrots are better quality than a typical wild tropical fruit, because they have less fiber and fewer toxic compounds. If the German raw-foodists had been eating wild foods, their energy balance and reproductive performance would have been much lower than found by Koebnick's team.

Supermarkets offer a year-round supply of the choicest foods, so the German raw-foodists had no seasonal shortages. Foragers, by contrast, cannot escape the tough times when sweet fruits, honey, or game meat become no more than occasional luxuries rather than daily pleasures. Even subsistence foods can then be hard to find. Anthropologist George Silberbauer reported that among the G/wi bushmen of the Central Kalahari, early summer was a time when all lost weight and everyone complained of hunger and thirst. In deserts like the Kalahari the result can be difficult indeed, but periodic shortages of energy like this are routine in all living hunter-gatherers, just as they are in rain-forest chimpanzees. Judging from studies of bones and teeth, which show in their fine structure the marks of nutritional stress, energy shortages were also universal in archaeological populations. Until the development of agriculture, it was the human fate to suffer regular periods of hunger—typically, it seems, for several weeks a year—even though they ate their food cooked.

Raw-foodism seems to be an increasingly popular habit, but if raw diets are so challenging, why do people like them? Raw-foodists are very enthusiastic about the health benefits, as described in books with such titles as *Self Healing Power! How to Tap Into the Great Power Within You.* They report a sense of well-being, better physical functioning, less bodily pain, more vitality, and improved emotional and social performance. There are claims of reductions in rheumatoid arthritis and fibromyalgia symptoms, less dental erosion, and improved antioxidant intake. Mostly such assertions have not been scientifically tested, but researchers have found improved serum cholesterol and triglyceride values.

Raw-foodists offer philosophical reasons too. "Natural nutrition is raw," asserted Stephen Arlin, Fouad Dini, and David Wolfe in *Nature's First Law*, a popular guide to raw-foodism. "It always has been. It always will be. . . . Cooked food is poison." Many follow the pseudoscientific ideas of vegetarian Edward Howell, who theorized in a 1946 book that plants contain "living" or "active" enzymes, which, if eaten raw, operate for our benefit inside our bodies. His followers therefore prepare their foods below a certain temperature, normally about 45–48°C (113–118°F), above which the "life force" of the enzymes is supposedly destroyed. To scientists the idea that food enzymes contribute to digestion or cellular function in our bodies is nonsense because these molecules are themselves digested in our stomachs and small intestines. The "living enzyme" idea also ignores that even if food enzymes survived our digestive systems, their own specific metabolic functions are too specialized to allow them to do anything useful in our bodies. But while the idea of a "life force" in "living foods" is not accepted by physiologists, it persuades many raw-foodists to persist in their diet. By permitting some use of low heat, Howell's philosophy also enables the "raw" food to be somewhat more palatable, easier to prepare, and more digestible than a truly unheated food would be.

Other raw-foodists are guided by moral principles. In 1813 the poet Percy Bysshe Shelley argued that meat eating was an appalling habit responsible for many of society's ills and was obviously unnatural, given that humans lack claws, have blunt teeth, and dislike raw meat. Since he concluded that the invention of cooking was responsible for meat eating, and hence for such problems as "tyranny, superstition, commerce, and inequality," he decided that humans were better off without cooking.

Instinctotherapists, a minority group among raw-foodists, believe that because we are closely related to apes we should model our eating behavior on theirs. In 2003 I had lunch with Roman Devivo and Antje Spors, whose book *Genefit Nutrition* argues that cooked food provides an unhealthy diet to which we are not adapted. They were lean and healthy. They were clear about their preference, which was to eat all their food not merely raw but without any preparation at all. They politely declined a salad because its ingredients had been chopped and mixed. The natural way, they explained, is to do what chimpanzees do. Just as those apes find only one kind of fruit when eating in a given tree, so we should eat only one kind of food in any meal.

To illustrate their habit, Devivo, Spors, and a friend had brought a basket containing a selection of organic foods. They sniffed at several fruits, one at a time, to allow their bodies to decide what would suit them best ("by instinct," they said). One chose apples; another chose a pineapple. Each ate only his or her first choice. The third decided on a protein-rich food. He had brought frozen buffalo steaks and pieces of buffalo femur. Today was a marrow day. The femur chunks were the size of golf balls. Inside each was a cold pink mush that looked like strawberry ice cream. He cleaned out several pieces of bone with a teaspoon.

However strange it may be to think that we should eat to conserve living enzymes, or to reduce violence, or in the manner of apes, such concepts are helpful to raw-foodists because they bolster a strong commitment to principle. Eating raw intrudes into social life, demands a lot of time in the kitchen, and requires a strong will to resist the thought of cooked food. It can create personal problems, such as annoyingly frequent urination, and for meat eaters it increases the risk of eating toxins or pathogens that would be destroyed by cooking. There are other health risks too. Recent studies indicate that low bone mass in the backs and hips of raw-foodists was caused by their raw diet. Raw diets are also associated with low levels of vitamin B12, low levels of HDL cholesterol (the "good" cholesterol), and elevated levels of homocysteine (a suspected risk factor for cardiovascular disease).

In theory the precarious energy budgets experienced by the Giessen study subjects could be misleading. Maybe modern raw-foodists are so far removed from nutritional wisdom that they are just not choosing the right combination of foods. What about reliance on raw food in nonindustrialized cultures? This has often been reported. At the end of the nineteenth century, anthropologist William McGee, president of the National Geographical Society and cofounder of the American Anthropological Association, claimed that the Seri hunter-gatherers of northwestern Mexico ate meat and carrion largely raw. Four thousand years ago Sumerians in the Third Dynasty of Ur said that the bedouin of the western desert ate their food raw. As late as 2007, pygmies in Uganda's Ruwenzori Mountains were reported in a national Ugandan newspaper to be living off raw food. Writers from Plutarch to colonial sailors of the nineteenth century made similar claims, but all have proved illusory, often colored by a racist tinge. "Only savages can be satisfied with the pure products of nature, eaten without seasoning and as nature provides them," sniffed the entry in an eighteenth-century encyclopedia. In 1870 anthropologist Edward Tylor examined all such accounts and found no evidence of any being real. He concluded that cooking was practiced by every known human society. Similarly, all around the world are societies that tell of their ancestors having lived without fire. When anthropologist James Frazer examined reports of prehistoric firelessness, he found them equally full of fantasy, such as fire being brought by a cockatoo or being tamed after it was discovered in a woman's genitals. The control of fire and the practice of cooking are human universals.

Still, in theory, societies could exist where cooked food is only a small part of the diet. The quirky nutritionist Howell thought so. In the 1940s he stated as part of his theory of the benefits of raw foods that the traditional Inuit (or Eskimo) diet was dominated by raw foods. His claim about the Inuit eating most of their food raw has been an important mainstay of the raw-foodist movement ever since.

But again it has proved exaggerated. The most detailed studies of un-Westernized Inuit diets were by Vilhjalmur Stefansson during a series of expeditions to the Copper Inuit beginning in 1906. Their diet was virtually plant-free, dominated by seal and caribou meat, supplemented by large salmonlike fish and occasional whale meat. Stefansson found that cooking was the nightly norm.

Every wife was expected to have a substantial meal ready for her husband when he got back from the hunt. In winter a husband came home at a predictably early time and would find the smell of boiling seal meat and steaming broth as soon as he entered the igloo. The long days of summer made the time of a husband's return home less predictable, so wives often went to bed before he came back. Anthropologist Diamond Jenness accompanied Stefansson, and described what happened if a wife failed to leave cooked meat for her husband: "Woe betide the wife who keeps him waiting after a day spent in fishing or hunting! . . . Her husband will probably beat her, or stamp her in the snow, and may even end by throwing her household goods after her and bidding her begone forever from his house."

Arctic cooking was difficult because of the shortage of fuel. In summer women made small twig fires, whereas in winter they cooked over burning seal oil or blubber in stone pots. After the snow had melted to water, the process of boiling meat took a further hour. Despite the difficulties, the meat was well cooked. "I have never seen Eskimo eat partly cooked meat so bloody as many steaks I have seen devoured in cities—when they cook, they usually cook well," Stefansson wrote in 1910.

The slow cooking and shortage of fuel meant it was hard for men to cook when they were out on the hunt, so during the day they would sometimes eat fresh fish raw, either the flesh or in the case of large fish, just the intestines. Hunters also made caches of excess fish, which they could recover later for a cold meal. However, even though these foods were uncooked they were affected by being stored: fish from the cache became "high"—in other words, smelly because they were partially rotten. Most people liked the strong taste. Jenness saw "a man take a bone from rotten caribou-meat cached more than a year before, crack it open and eat the marrow with evident relish although it swarmed with maggots."

Though many raw foods were eaten for convenience, some were taken by choice. Blubber was often preferred raw. It was soft and could be spread easily over meat like butter. Other meats eaten raw were also soft, such as seal livers and kidneys and caribou livers. Occasionally there was evidence of more exotic tastes. Stefansson's hosts were horrified to hear of a distant group, the Puiplirmiut, who supposedly collected frozen deer droppings off the snow and ate them like berries. They said that was a truly repulsive habit, and anyway it was a waste of a good dropping. Those pellets were a fine food, they said, when boiled and used to thicken blood soup. The only vegetable food that was regularly eaten raw was the lichen eaten by caribou, which the Copper Inuit ate when the lichen was partially digested. In summer they would take it directly from the rumen and eat it while cutting up the carcass. As the cold closed in during the fall, they were more likely to allow the full stomach to freeze intact with the lichens inside. They would then cut it into slices for a frozen treat.

The Inuit probably ate more raw animal products than other societies, but like every culture the main meal of the day was taken in the evening, and it was cooked. In a scene captured by anthropologist Jiro Tanaka, the !Kung of the Kalahari illustrate the typical pattern for hunter-gatherers of a light breakfast and snacks during the day, followed by an evening meal. "Finally, as the sun begins to set, each woman builds a large cooking fire near her hut and commences cooking.... The hunters return to camp in the semidarkness, and each family eats supper around the fire after darkness has fallen.... Only in the evening does the whole family gather to eat a solid meal, and indeed people consume the greater part of their daily food then. The only exception is after a big kill, when a large quantity of meat has been brought back to camp: then people eat any number of times during the day, keeping their stomachs full to bursting, until all the meat is gone."

The Inuit consumed raw food mostly as a snack out of camp, as is typical of human foragers. In 1987, anthropologist Jennifer Isaacs described which foods Australian aborigines ate raw or cooked. Although foragers sometimes lit fires in the bush to cook quick meals such as mud crabs (a particular favorite), the majority of animal items were brought back to camp to be cooked. A few items, such as a species of mangrove worm, were always eaten raw, and these were not brought back to camp. Isaacs reported three types of food that were eaten sometimes raw and sometimes cooked—turtle eggs, oysters, and witchetty grubs—and in each case they were eaten raw by people foraging far from camp but were cooked if eaten in camp. Most fruits are preferred raw and are eaten in the bush, whereas roots, seeds, and nuts are brought back to camp to be cooked. Everywhere we look, home cooking is the norm. For most foods, eating raw appears to be a poor alternative demanded by circumstance.

What happens to people who are forced to eat raw diets in wild habitats, such as lost explorers, castaways, or isolated adventurers simply trying to survive despite losing their ability to cook? This category of people offers a third test of how well humans can utilize raw food. You might think that when humans are forced to eat raw, they would grumble at the loss of flavor but nevertheless be fine. However, I have not been able to find any reports of people living long term on raw wild food.

The longest case that I found of survival on raw animal foods lasted only a few weeks. In 1972 a British sailor, Dougal Robertson, and his family lost their boat to killer whales in the Pacific and were confined to a dinghy for thirty-eight days. They began with a few cookies, oranges, and glucose candies. By the seventh day they were forced to eat what they could catch on a line. They spent their last thirty-one days at sea mostly eating raw turtle meat, turtle eggs, and fish. There were occasional treats, such as chewing the liver and heart of a shark, but their staple was a "soup" of dried turtle in a mix of rainwater, meat juice, and eggs.

They caught more food than they could eat, and they survived in good cheer. Indeed, their diet suited them so well that by the end of their ordeal, Robertson reported that their physical condition was actually better than when they had begun their journey. Sores that had been present when their boat was sunk had healed, and their bodies were functioning effectively. The only problem was that nine-year-old Neil, despite being given extra portions of bone marrow, was disturbingly thin.

And all were hungry. They "enjoyed the flavour of the raw food as only starving people can." Their fantasies focused on cooked food. By the twenty-fourth day, Robertson recorded, "our daydreams had switched from ice cream and fruit to hot stews, porridge, steak and kidney puddings, hotpots and casseroles. The dishes steamed fragrantly in our imaginations and as we described their smallest details to each other we almost tasted the succulent gravies as we chewed our meager rations." The Robertsons' raw diet supported survival but it also brought a sense of starvation.

Their resourcefulness enabled them to emerge from a terrifying situation in fine condition. They may have been hungry and thinner, but they were apparently not starving to the point of danger. Their experience shows that with abundant food, people can survive well on a raw animal-based diet for at least a month. But people sometimes survive with no food at all for a month, provided they have water. The lack of any evidence for longer-term survival on raw wild food suggests that even *in extremis*, people need their food cooked.

The case that comes closest to long-term survival on raw wild food is that of Helena Valero. This exceptional woman was a Brazilian of European descent who reportedly survived in a remote forest for some seven months in the 1930s. She knew the jungle well because at about age twelve she had been kidnapped by Yanomamö Indians. She became a member of their tribe but her experience was very hard. One day, after her life was threatened, she escaped her captors. She took a firebrand wrapped in leaves so she could cook, but after a few days a heavy rain drenched it. Unwilling to return to Yanomamö life, she wandered alone, fireless and increasingly hungry, until she found an abandoned banana plantation. Valero was lucky because villagers had planted the trees in a dense grove. There, she said, she survived by eating raw bananas. She counted the seven months by the passage of the moon. Valero did not record her condition at the end of her exile, but she was eventually found by Yanomamö. She returned to the comforts of village life, married twice, had four children, and eventually feared for her children's lives and escaped again at about age thirty-five. She never found happiness in Brazilian society.

Valero's tale could not be verified, but if anyone were to survive on raw food in the wild, it makes sense that they would have the fortune to have an abundant supply of a high-calorie domesticated fruit. Bananas are often touted as nature's most perfect food.

In more ordinary circumstances starvation is a rapid threat when eating raw in the wild. Anthropologist Allan Holmberg was at a remote mission station in Bolivia in the 1940s when a group of seven Siriono hunter-gatherers arrived from the forest. They were so hungry and emaciated that, as one of them told Holmberg, if they had not arrived when they did they might have died. This group had been part of a band that had thrived in the rain forest until they were taken to a government school. They had been so resentful of their forced removal that they had escaped with the aim of returning to their ancestral homeland. To avoid capture they had moved fast, walking even in heavy rain. Without proper cover the smoldering logs they were carrying were extinguished. After that the little group was reduced to a raw diet of wild plants until they were rescued after three weeks. They walked less than five miles per day and even though they knew the forest intimately and found raw plants to eat, they still could not obtain sufficient energy from their diets. Two of the men had bows and there was lots of game, so they might have done better but for a taboo on raw meat, which they claimed not to eat under any conditions. But even hunter-gatherers often live well with little meat for weeks on end, as long as they cook. The Siriono experience suggests that raw diets are dangerous because they do not provide enough energy.

In 1860 Robert Burke and William Wills led an ill-fated expedition from southern to northern Australia. When they ran out of food they asked the local Yandruwandha aborigines for help. The Yandruwandha were living on the abundant nardoo plant. They pounded nardoo seeds into a bitter flour, washed it, and then cooked it. The explorers liked the flour but apparently omitted the washing and cooking. The result was disaster. "I am weaker than ever," wrote Wills, "although I have a good appetite, and relish the nardoo much, but it seems to give us no nutriment." Burke and Wills died from poisoning, starvation, or both. However, they had a companion who survived and joined the Yandruwandha, ate lots of cooked nardoo flour, and was in excellent condition when he was rescued ten weeks later.

The cases I have listed are exceptional because it is rare for people to even attempt to survive on raw food in the wild. When Thor Heyerdahl took a primitive raft across the Pacific to test his theories about prehistoric migrations, he had a primus stove with him and one of his crew was a cook. When an airplane crashed in the Chilean Andes in 1972 and stranded twenty-seven people for seventy-one days, the survivors resorted to cannibalism and cooked the meat. When the whale ship *Essex* went down in the Pacific and its sailors cannibalized one another in small lifeboats, they cooked on stone-bottomed fires. Several Japanese soldiers lived alone in the jungle after World War II. One of them, Shoichi Yokio, stayed in Guam until 1972, surviving on fruits, snails, eels, and rats. But he did not eat them raw. Life in his underground cave depended on his smoke-blackened pots, just as it did for all such holdouts.

Perhaps the most famous real-life castaway was Alexander Selkirk, the model for Robinson Crusoe. In 1704, after quarrelling with the captain of his ship and rashly demanding to be put ashore, Selkirk began more than four years alone on the island of Más a Tierra, 670 kilometers (416 miles) west of Chile in the Pacific Ocean. He had his Bible, a musket with a pound of powder, some mathematical instruments, a hatchet, a knife, and a few carpenter's tools. He ended up very wild, dancing with his tamed goats and cats and barely recognizable as human. But when his gunpowder was nearly spent, "he got fire by rubbing two sticks of Piemento Wood together upon his knee." He was able to cook throughout his time in isolation.

Raw-foodists, it is clear, do not fare well. They thrive only in rich modern environments where they depend on eating exceptionally high-quality foods. Animals do not have the same constraints: they flourish on wild raw foods. The suspicion prompted by the shortcomings of the Evo Diet is correct, and the implication is clear: there is something odd about us. We are not like other animals. In most circumstances, we need cooked food.

CHAPTER 2

The Cook's Body

"Domestication of fire probably reacted on man's physical development as well as on his culture, for it would have reduced some selective pressures and increased others. As cooked food replaced a diet consisting entirely of raw meat and fresh vegetable matter, the whole pattern of mastication, digestion, and nutrition was altered."

—KENNETH OAKLEY, *Social Life of Early Man*

Although humans fare poorly on raw diets nowadays, at some time our ancestors must have utilized bush fruits, fresh greens, raw meat, and other natural products as efficiently as apes do. What can account for the change? Why, given all the obvious advantages of being able to extract large amounts of energy from raw food, have humans lost this ancient ability?

In theory an evolutionary mishap might be responsible for this failure of our biology: the genetic coding for a well-adapted digestive system could have been lost by chance. But a failure of evolutionary adaptation is an unlikely explanation for something as widespread and labor-intensive as cooking. Natural selection mostly generates exquisitely successful designs, particularly for features that are as important and in such regular use as our intestinal systems. We can expect to find a compensatory benefit that has been made possible by our inability to utilize raw food effectively.

Evolutionary trade-offs are common. Compared to chimpanzees, we climb badly but we walk well. Our awkwardness in trees is due partly to our having long legs and flat feet, but those same legs and feet enable us to walk more efficiently than other apes. In a similar way, our limited effectiveness at digesting raw food is due to our having relatively small digestive systems compared to those of our cousin apes. But the reduced size of our digestive systems, it seems, enables us to process cooked food with exceptional proficiency.

We can think of cooked food offering two kinds of advantage, depending on whether species have adapted to a cooked diet. Spontaneous benefits are experienced by almost any species, regardless of its evolutionary history, because cooked food is easier to digest than raw food. Domestic animals such as calves, lambs, and piglets grow faster when their food is cooked, and cows produce more fat in their milk and more milk per day when eating cooked rather than raw seeds. A similar effect appears in fish farms. Salmon grow better on a diet of cooked rather than raw fishmeal. No wonder farmers like to give cooked mash or swill to their livestock. Cooked food promotes efficient growth.

The spontaneous benefits of cooked food explain why domesticated pets easily become fat: their food is cooked, such as the commercially produced kibbles, pellets, and nuggets given to dogs and cats. Owners of obese pets who recognize this connection and see cooked food as a health threat sometimes choose to feed raw food to their beloved ones to help them lose weight. Biologically Appropriate Raw Food, or BARF, is a special diet advertised as being beneficial for dogs for the same reason that raw-foodists advocate raw diets for humans: it is natural. "Every living animal on earth requires a biologically appropriate diet. And if you think about it, not one animal on earth is adapted by evolution to eat a cooked food diet. This means the BARF diet is exactly what we should be feeding our pets." The effects of this diet is reminiscent of raw-foodists' experience: "You can always tell a raw-food dog; they look better, have more energy, are thin and vibrant," says an owner of a golden retriever whose coat started glowing within a week of eating raw food exclusively.

Even insects appear to get the spontaneous benefits of cooked food. Researchers rearing agricultural pests in large numbers to find out how to control them give each insect species its own particular recipe of cooked food. Larvae of the diamondback moth thrive on a toasted mix of wheat germ, casein, bean meal, and cabbage flour. Black vine weevils do best on thoroughly boiled and blended lima beans. Whether domestic or wild, mammal or insect, useful or pest, animals adapted to raw diets tend to fare better on cooked food.

In humans, because we have adapted to cooked food, its spontaneous advantages are complemented by evolutionary benefits. The evolutionary benefits stem from the fact that digestion is a costly process that can account for a high proportion of an individual's energy budget—often as much as locomotion does. After our ancestors started eating cooked food every day, natural selection favored those with small guts, because they were able to digest their food well, but at a lower cost than before. The result was increased energetic efficiency.

Evolutionary benefits of adapting to cooked food are evident from comparing human digestive systems with those of chimpanzees and other apes. The main differences all involve humans having relatively small features. We have small mouths, weak jaws, small teeth, small stomachs, small colons, and small guts overall. In the past, the unusual size of these body parts has mostly been attributed to the evolutionary effects of our eating meat, but the design of the human digestive system is better explained as an adaptation to eating cooked food than it is to eating raw meat.

Mick Jagger's biggest yawn is nothing compared to a chimpanzee's. Given that the mouth is the entry to the gut, humans have an astonishingly tiny opening for such a large species. All great apes have a prominent snout and a wide grin: chimpanzees can open their mouths twice as far as humans, as they regularly do when eating. If a playful chimpanzee ever kisses you, you will never forget this point. To find a primate with as relatively small an aperture as that of humans, you have to go to a diminutive species, such as a squirrel monkey, weighing less than 1.4 kilograms (3 pounds). In addition to having a small gape, our mouths have a relatively small volume—about the same size as chimpanzee mouths, even though we weigh some 50 percent more than they do. Zoologists often try to capture the essence of our species with such phrases as the naked, bipedal, or big-brained ape. They could equally well call us the small-mouthed ape.

The difference in mouth size is even more obvious when we take the lips into account. The amount of food a chimpanzee can hold in its mouth far exceeds what humans can do because, in addition to their wide gape and big mouths, chimpanzees have enormous and very muscular lips. When eating juicy foods like fruits or meat, chimpanzees use their lips to hold a large wad of food in the outer part of their mouths and squeeze it hard against their teeth, which they may do repeatedly for many minutes before swallowing. The strong lips are probably an adaptation for eating fruits, because fruit bats have similarly large and muscular lips that they use in the same way to squeeze fruit wads against their teeth. Humans have relatively tiny lips, appropriate for a small amount of food in the mouth at one time.

Our second digestive specialization is having weaker jaws. You can feel for yourself that our chewing muscles, the temporalis and masseter, are small. In nonhuman apes these muscles often reach all the way from the jaw to the top of the skull, where they sometimes attach to a ridge of bone called the sagittal crest, whose only function is to accommodate the jaw muscles. In humans, by contrast, our jaw muscles normally reach barely halfway up the side of our heads. If you clench and unclench your teeth and feel the side of your head, you have a good chance of being able to prove to yourself that you are not a gorilla: your temporalis muscle likely stops near the top of your ear. We also have diminutive muscle fibers in our jaws, one-eighth the size of those in macaques. The cause of our weak jaws is a human-specific mutation in a gene responsible for producing the muscle protein myosin. Sometime around two and a half million years ago this gene, called MYH16, is thought to have spread throughout our ancestors and left our lineage with muscles that have subsequently been uniquely weak. Our small, weak jaw muscles are not adapted for chewing tough raw food, but they work well for soft, cooked food.

Human chewing teeth, or molars, also are small—the smallest of any primate species in relation to body size. Again, the predictable physical changes in food that are associated with cooking account readily for our weak chewing and small teeth. Even without genetic evolution, animals reared experimentally on soft diets develop smaller jaws and teeth. The reduction in tooth size produces a well-adapted system: physical anthropologist Peter Lucas has calculated that the size of a tooth needed to make a crack in a cooked potato is 56 percent to 82 percent smaller than needed for a raw potato.

Continuing farther into the body, our stomachs again are comparatively small. In humans the surface area of the stomach is less than one-third the size expected for a typical mammal of our body weight, and smaller than in 97 percent of other primates. The high caloric density of cooked food suggests that our stomachs can afford to be small. Great apes eat perhaps twice as much by weight per day as we do because their foods are packed with indigestible fiber (around 30 percent by weight, compared to 5 percent to 10 percent or less in human diets). Thanks to the high caloric density of cooked food, we have modest needs that are adequately served by our small stomachs.

Below the stomach, the human small intestine is only a little smaller than expected from the size of our bodies, reflecting that this organ is the main site of digestion and absorption, and humans have the same basal metabolic rate as other primates in relation to body weight. But the large intestine, or colon, is less than 60 percent of the mass that would be expected for a primate of our body weight. The colon is where our intestinal flora ferment plant fiber, producing fatty acids that are absorbed into the body and used for energy. That the colon is relatively small in humans means we cannot retain as much fiber as the great apes can and therefore cannot utilize plant fiber as effectively for food. But that matters little. The high caloric density of cooked food means that normally we do not need the large fermenting potential that apes rely on.

Finally, the volume of the entire human gut, comprising stomach, small intestine, and large intestine, is also relatively small, less than in any other primate measured so far. The weight of our guts is estimated at about 60 percent of what is expected for a primate of our size: the human digestive system as a whole is much smaller than would be predicted on the basis of size relations in primates.

Our small mouths, teeth, and guts fit well with the softness, high caloric density, low fiber content, and high digestibility of cooked food. The reduction increases efficiency and saves us from wasting unnecessary metabolic costs on features whose only purpose would be to allow us to digest large amounts of high-fiber food. Mouths and teeth do not need to be large to chew soft, high-density food, and a reduction in the size of jaw muscles may help us produce the low forces appropriate to eating a cooked diet. The smaller scale may reduce tooth damage and subsequent disease. In the case of intestines, physical anthropologists Leslie Aiello and Peter Wheeler reported that compared to that of great apes, the reduction in human gut size saves humans at least 10 percent of daily energy expenditure: the more gut tissue in the body, the more energy must be spent on its metabolism. Thanks to cooking, very high-fiber food of a type eaten by great apes is no longer a useful part of our diet. The suite of changes in the human digestive system makes sense.

Could the tight fit between the design of our digestive systems and the nature of cooked food be deceptive? The character Pangloss in Voltaire's *Candide* claimed that our noses were designed to carry spectacles, based on the fact that our noses support spectacles efficiently. But actually spectacles have been designed to fit on noses, rather than the other way around. Following Pangloss's reasoning, in theory cooked food might similarly be well suited for a human gut that had been adapted for another kind of diet.

Meat is the obvious possibility. The "Man-the-Hunter" hypothesis assumes our ancestors were originally plant eaters, with the last species to eat relatively little meat being the australopithecine that gave rise to habilines more than two million years ago. Much of the australopithecines' plant food would have had the low caloric density and high fiber concentration seen in great-ape diets. We should therefore expect those ancient apes to have had large digestive systems, as chimpanzees and gorillas do today. In support of this idea, fossils show that australopithecines had broad hips and a rib cage that was flared outward toward the waist. Both features indicate the presence of capacious guts, held by the rib cage and supported by the pelvis. According to the meat-eating scenario, as increased amounts of meat were eaten by habilines and their descendants, modifications must have evolved in the mouth and digestive system.

Physical anthropologist Peter Ungar reported in 2004 that the molars (chewing teeth) of very early humans were somewhat sharper than those of their australopithecine ancestors. They might therefore have been adapted to eating tough foods, including raw meat. Carnivores such as dogs, and probably wolves and hyenas, also tend to have small guts compared to those of great apes, including small colons that are efficient for the high caloric density and low fiber density of a meat diet. But despite these hints of humans being designed for meat eating, our mouths, teeth, and jaws are clearly not well adapted to eating meat unless it has been cooked. Raw wild meat from game animals is tough, which is partly why cooking is so important. Advocates of the meat-eating hypothesis have themselves noted that humans differ from carnivores by our having small mouths, weak jaws, and small teeth that cannot easily shear flesh.

The way food moves through our bodies compounds the problem. In carnivores, meat spends a long time in the stomach, allowing intense muscular contractions of the stomach walls to reduce raw meat to small particles that can be digested rapidly. Dogs tend to keep food in the stomach for two to four hours, and cats for five to six hours, before passing the food quickly through the small intestine. By contrast, humans resemble other primates in keeping food in our stomachs for a short time, generally one to two hours, and then passing it slowly through the small intestine. Lacking the carnivore system of retaining food for many hours in our stomachs, we humans are inefficient at processing chunks of raw meat.

If our mouths, teeth, jaws, and stomachs all indicate that humans are not adapted to eating lumps of raw meat, they might in theory be designed to digest meat that has been processed without being cooked. Raw meat might have been usefully pounded to make it easily chewed. It might have been allowed to rot, in parts of the world that were sufficiently cold for bacterial infection not to be a major threat. Or it might have been dried. But these ideas cannot solve the problem of how plant foods were eaten.

The problem is that tropical hunter-gatherers have to eat at least half of their diet in the form of plants, and the kinds of plant foods our hunter-gatherer ancestors would have relied on are not easily digested raw. So if the meat-eating hypothesis is advanced to explain why *Homo erectus* had small teeth and guts, it faces a difficulty with the plant component of the diet. It cannot explain how a human with a diminished capacity for digestion could have digested plant foods efficiently.

Plants are a vital food because humans need large amounts of either carbohydrates (from plant foods) or fat (found in a few animal foods). Without carbohydrates or fat, people depend on protein for their energy, and excessive protein induces a form of poisoning. Symptoms of protein poisoning include toxic levels of ammonia in the blood, damage to the liver and kidneys, dehydration, loss of appetite, and ultimately death. The grim result was described by Vilhjalmur Stefansson based on his experience in the

Arctic in a lean season when fat was so scarce (and plant foods were absent, as usual) that protein became the predominant macronutrient in the diet. "If you are transferred suddenly from a diet normal in fat to one consisting wholly of . . . [lean meat] you eat bigger and bigger meals for the first few days until at the end of about a week you are eating in pounds three or four times as much as you were at the beginning of the week. By that time you are showing both signs of starvation and protein poisoning. You eat numerous meals; you feel hungry at the end of each; you are in discomfort through distension of the stomach with much food and you begin to feel a vague restlessness. Diarrhoea will start in from a week to 10 days and will not be relieved unless you secure fat. Death will result after several weeks."

Because the maximum safe level of protein intake for humans is around 50 percent of total calories, the rest must come from fat, such as blubber, or carbohydrates, such as in fruits and roots. Fat is an excellent source of calories in high-latitude sites like the Arctic or Tierra del Fuego, where sea mammals have evolved thick layers of blubber to protect themselves from the cold. However, fat levels are much lower in the meat of tropical mammals, averaging around 4 percent, and high-fat tissues like marrow and brain are always in limited supply. The critical extra calories for our equatorial ancestors therefore must have come from plants, which are vital for all tropical hunter-gatherers. During periods of food shortage, such as the annual dry seasons, fat levels in meat would have been particularly low, down to 1 percent to

2 percent. A carbohydrate supply from plant foods would then have been especially vital.

But if early humans had the same small guts as we do, they could not have obtained their plant carbohydrates without cooking. Recall the poor metabolic performance of the urban raw-foodists in the Giessen study. Those people ate very high-quality cultivated food processed with the aid of sprouting, blending, and even low-temperature ovens, yet still obtained so little energy that reproductive function was seriously impaired. If our early human ancestors indeed ate their plant food raw, they would have needed to find ways of processing it that were superior to our modern technology. But it is not credible that Stone Age people developed non-thermal methods of food preparation more effective than using an electric blender.

Hunter-gatherers living on raw food might sometimes have found plant foods of an exceptionally high caloric density, such as avocados, olives, or walnuts. But no modern habitats produce such foods in abundance all year. Perhaps a few lost places would have had highly productive natural orchards until they were replaced by agriculture, such as the fertile valleys of the Middle East. But occasional productive areas would not explain the wide geographical range of human ancestors across Africa, Europe, and Asia by 1.8 million years ago. Furthermore, seasonal scarcities occur in every habitat and would have forced people to use foods of lower caloric density, such as roots. The notion of a permanently superproductive habitat is unrealistic. People with an anatomy like ours today could not have flourished on raw food in the Pleistocene epoch.

Beyond reducing the size of teeth and guts, the adoption of cooking must have had numerous effects on our digestive system because it changed the chemistry of our food. Cooking would have created some toxins, reduced others, and probably favored adjustments to our digestive enzymes. Very little is known about how our detoxification system and enzyme chemistry differ from those of great apes, but studies should eventually provide further tests of the hypothesis that human bodies are adapted to eating cooked foods.

Take, for example, Maillard compounds, such as heterocyclic amines and acrylamide. These complex molecules are formed from a process that begins with the union of sugars and amino acids, particularly lysine. Maillard compounds occur naturally in our bodies and increase in frequency with age. They occur at low concentration in natural foods but under the influence of heat their concentration becomes much higher than what is found in nature, whether in smoke (from fires or cigarettes) or cooked items. Their presence is easily recognized in the brown colors found in pork crackling or bread crust. Maillard compounds cause mutations in bacteria and are suspected of leading to some human cancers. They can also induce a chronic state of inflammation, a process that raw-foodists invoke to explain why they feel better on raw diets. The cooking hypothesis suggests that our long evolutionary history of exposure to Maillard compounds has led humans to be more resistant to their damaging effects than other mammals are. It is an important question because many processed foods contain Maillard compounds that are known to cause cancer in other animals. Acrylamide is an example. In 2002 acrylamide was discovered to occur widely in commercially produced potato products, such as potato chips. If it is as carcinogenic to humans as it is to other animals, it is dangerous. If not, it may provide evidence of human adaptation to Maillard compounds, and hence of a long exposure to heated foods.

Evolutionary adaptation to cooking might likewise explain why humans seem less prepared to tolerate toxins than do other apes. In my experience of sampling many wild foods eaten by primates, items eaten by chimpanzees in the wild taste better than foods eaten by monkeys. Even so, some of the fruits, seeds, and leaves that chimpanzees select taste so foul that I can barely swallow them. The tastes are strong and rich, excellent indicators of the presence of non-nutritional compounds, many of which are likely to be toxic to humans—but presumably much less so to chimpanzees.

Consider the plum-size fruit of *Warburgia ugandensis*, a tree famous for its medicinal bark. *Warburgia* fruits contain a spicy compound reminiscent of a mustard oil. The hot taste renders even a single fruit impossibly unpleasant for humans to ingest. But chimpanzees can eat a pile of these fruits and then look eagerly for more.

Many other fruits in the chimpanzee diet are almost equally unpleasant to the human palate. Astringency, the drying sensation produced by tannins and a few other compounds, is common in fruits eaten by chimpanzees. Astringency is caused by the presence of tannins, which bind to proteins and cause them to precipitate. Our mouths are normally lubricated by mucoproteins in our saliva, but because a high density of tannins precipitates those proteins, it leaves our tongues and mouths dry: hence the "furry" sensation in our mouths after eating an unripe apple or drinking a tannin-rich wine. One has the same experience when tasting chimpanzee fruits such as *Mimusops bagshawei* or the widespread *Pseudospondias microcarpa.* Though chimpanzees can eat more than 1 kilogram (2.2 pounds) of such fruits during an hour or more of continuous chewing, we cannot. Some other chimpanzee foods taste bitter to us, such as certain figs. Still other fruits elicit in us unusual sensations, such as the fruits of *Monodora myristica,* whose sharp and lemony taste is followed by a numbing sensation at the tip of the tongue like that caused by novocaine. Of the scores of chimpanzee foods I have tasted, I could imagine filling my belly with only a very few species, such as a wild raspberry—but alas, one rarely finds more than a handful of these delicious fruits at a time. The shifts in food preference between chimpanzees and humans suggest that our species has a reduced physiological tolerance for foods high in toxins or tannins. Since cooking predictably destroys many toxins, we may have evolved a relatively sensitive palate.

By contrast, if we were adapted to a raw-meat diet we would expect to see evidence of resistance to the toxins produced by bacteria that live on meat. No such evidence is known. Even when we cook our meat we are vulnerable to bacterial infections. The U.S. Centers for Disease Control and Prevention state that at least forty thousand cases of food poisoning by *Salmonella* alone are reported annually in the United States, and as many as one million cases may go unreported. The estimated total number of cases due to the top twenty harmful bacteria, including *Staphylococcus, Clostridium, Campylobacter, Listeria, Vibrio, Bacillus,* and *Escherichia coli* (*E. coli*), is in the tens of millions per year. The best prevention is to cook meat, fish, and eggs beyond 140°F (60°C), and not to eat foods containing unpasteurized milk or eggs. The cooking hypothesis suggests that because our ancestors have typically been able to cook their meat, humans have remained vulnerable to bacteria that live on raw meat.

Anthropology has traditionally adopted the Man-the-Hunter scenario, proposing our species as a creature that was modified from australopithecines principally by our tendency to eat more meat. Certainly meat eating has been an important factor in human evolution and nutrition, but it has had less impact on our bodies than cooked food. We fare poorly on raw diets, no cultures rely on them, and adaptations in our bodies explain why we cannot easily utilize raw foods. Even vegetarians thrive on cooked diets. We are cooks more than carnivores. No wonder raw-foodism is a good way to lose weight.

CHAPTER 3

The Energy Theory of Cooking

"A man does not live on what he eats, an old proverb says, but on what he digests."

—JEAN ANTHELME BRILLAT-SAVARIN, *The Physiology of Taste: Or Meditations on Transcendental Gastronomy*

An obvious implication of animals and humans gaining more weight and reproducing better on cooked than raw diets is that when a food is heated, it must yield more energy. Yet authoritative science flatly challenges this idea. The U.S. Department of Agriculture's *National Nutrient Database for Standard Reference* and Robert McCance and Elsie Widdowson's *The Composition of Foods* are the principal sources for public understanding of the nutrient data for thousands of foods in the United States and the United Kingdom, respectively. They provide the data for our food labels. These references report that the effect of cooking on energy content is the same for beef, pork, chicken, duck, beetroot, potatoes, rice, oats, pastries, and dozens of other foods—on average, zero. According to these and similar compilations, cooking has important effects in changing water content and reducing the concentration of vitamins, but the density of calories supposedly remains unchanged whether food is eaten raw or is roasted, grilled, or boiled.

This conclusion is very puzzling. Obviously it conflicts with the abundant evidence that humans and animals get more energy from cooked foods. It also conflicts with various contrary conclusions from nutritional science. On the one hand, a widespread idea states that cooking is "a technological way of externalizing part of the digestive process," a claim that seems to imply some kind of benefit such as accelerated digestion. On the other hand, cooking is sometimes claimed to have a negative effect on energy value. I recently spotted some small "fresh premium breakfast sausages" in my local supermarket. The food label gave their energy content in calories. With a curious nod to those who might want to eat raw sausages, it included values for both the raw and the cooked product. "Serving size 2 links. Raw 130 cals (60 from fat). Cooked 120 cals (60 from fat)." The claim might seem surprising, but cooking can reduce calories in various ways. Cooking can lead to the loss of nutrient-filled juices. It can generate indigestible molecules such as Maillard compounds, reducing the amount of sugar or amino acids available for digestion. It can burn carbohydrates. It can lead to changes in texture that reduce a food's digestibility. Leading nutritionist David Jenkins judged such effects significant:

"The predominant effect (of cooking) is . . . to reduce the digestibility of the proteins."

Although different nutritionists say that cooking has no effect on the caloric content of food, or increases it, or decreases it, we can clear up this confusion. As indicated by the evidence from raw-foodists and the immediate benefits experienced by many animals eating cooked food, I believe the effects of cooking on energy gain are consistently positive. The mechanisms increasing energy gain in cooked food compared to raw food are reasonably well understood. Most important, cooking gelatinizes starch, denatures protein, and softens everything. As a result of these and other processes, cooking substantially increases the amount of energy we obtain from our food.

Starchy foods are the key ingredient of many familiar items such as breads, cakes, and pasta. They constitute almost all the world's major plant staples. In 1988–1990, cereals such as rice and wheat made up 44 percent of the world's food production, and together with just a few other starchy foods (roots, tubers, plantains, and dry pulses) accounted for 63 percent of the average diet. Starchy foods make up more than half of the diets of tropical hunter-gatherers today and may well have been eaten in similar quantity by our human and pre-human ancestors in the African savannas.

The most direct studies of the impact of cooking measure digestibility, meaning the proportion of a food our bodies digest and absorb. If the digestibility of a particular kind of starch is 100 percent, the starch is a perfect food: every part of it is converted into useful food molecules. If it is zero percent, the starch is completely resistant to digestion and provides no food value at all. The question is, how much does cooking affect the digestibility of starchy foods?

Our digestive system consists of two distinct processes. The first is digestion by our own bodies, which starts in the mouth, continues in the stomach, and is mostly carried out in the small intestine. The second is digestion, or strictly fermentation, by four hundred or more species of bacteria and protozoa in our large intestine, also known as the colon or large bowel. Foods that are digested by our bodies (from the mouth to the small intestine) produce calories that are wholly useful to us. But those that are digested by our intestinal flora yield only a fraction of their available energy to us—about half in the case of carbohydrates such as starch, and none at all in the case of protein.

This two-part structure means that the only way to assess how much energy a food provides is to calculate ileal digestibility, which samples the intestinal contents at the end of the small intestine, or ileum. The procedure requires scientists to conduct research on ileostomy patients, who have had their large intestine surgically removed and have a bag, or stoma, where the ileum ends. Researchers collect the ileal effluent from this bag.

Studies of ileal digestibility show that we use cooked starch very efficiently. The percentage of cooked starch that has been digested by the time it reaches the end of the ileum is at least 95 percent in oats, wheat, potatoes, plantains, bananas, cornflakes, white bread, and the typical European or American diet (a mixture of starchy foods, dairy products, and meat). A few foods have lower digestibility: starch in home-cooked kidney beans and flaked barley has an ileal digestibility of only around 84 percent.

Comparable measurements of the ileal digestibility of raw starch are much lower. Ileal digestibility is 71 percent for wheat starch, 51 percent for potatoes, and a measly 48 percent for raw starch in plantains and cooking bananas. The differences conform to test-tube studies of a wide range of items showing that raw starch is poorly digested, often only half as well as cooked starch. Starch granules eaten raw frequently pass through the ileum whole and enter the colon unchanged from when they were eaten. This "resistant starch" is vivid testimony to the deficits of a raw starch diet, explaining why we like our starch cooked and contributing to the weight loss that raw-foodists experience.

The principal way cooking achieves its increased digestibility is by gelatinization. Starch inside plant cells comes as dense little packages of stored glucose called granules. The granules are less than a tenth of a millimeter (four-thousandths of an inch) long, too small to be seen with the naked eye or to be damaged by the milling of flour, and they are so stable that in a dry environment they can persist for tens of thousands of years. However, as starch granules are warmed up in the presence of water they start to swell—at around

58°C (136°F) in the case of wheat starch, a well-studied and representative example. The granules swell because hydrogen bonds in the glucose polymers weaken when they are exposed to heat, and this causes the tight crystalline structure to loosen. By 90°C (194°F), still below boiling, the granules are disrupted into fragments. At this point the glucose chains are unprotected, and gelatinize. Starch does not necessarily stay gelatinized after being cooked. In day-old bread the starch reverts and becomes resistant. This might help explain why we like to toast bread after it has lost its initial freshness.

Gelatinization happens whenever starch is cooked, whether in the baking of bread, the gelling of pie fillings, the production of pasta, the fabrication of starch-based snack foods, the thickening of sauces, or, we can surmise, the tossing of a wild root onto a fire. As long as water is present, even from the dampness of a fresh plant, the more that starch is cooked, the more it is gelatinized. The more starch is gelatinized, the more easily enzymes can reach it, and therefore the more completely it is digested. Thus cooked starch yields more energy than raw.

This effect is detected easily in blood measurements. Within thirty minutes of a person eating a test meal of pure glucose, the concentration of glucose in his or her blood rises dramatically, before returning to base levels in just over an hour. The effect of eating cornstarch is almost identical as long as it is cooked. But following a meal of raw cornstarch, the value of blood glucose remains persistently low, peaking at less than a third of the value for cooked cornstarch.

The effects of cooking are captured by comparing the glycemic index of cooked and raw foods. Glycemic index (GI) is a widely used nutritional measure of a food's effect on blood sugar levels. High-GI foods, such as pure sugar, white bread, and potatoes, are good sources of energy after exercise, but for most people they are poor foods because they easily lead to excessive weight gain. In addition, the calories they offer tend to be "empty," being low in protein, essential fatty acids, vitamins, and minerals. Low-GI foods, such as whole-grain bread, high-fiber cereals, and vegetables, reduce weight gain, improve diabetes control, and lower cholesterol. Cooking consistently increases the glycemic index of starchy foods.

Animal protein has been almost as important as starch in diets throughout our evolution, and it remains a strongly preferred food today. Yet the effects of cooking on the energy derived from eating meat have never been formally investigated, particularly the effects due to meat's complex structure. Even the effects on proteins are a matter of debate. Until recently some scientists, such as David Jenkins, saw cooking as reducing protein digestibility. Others claim cooking protein is beneficial or has no effect. Recent studies of the digestion of eggs are starting to resolve the argument, showing for the first time that cooked protein is digested much more completely than raw protein.

In contrast to the new finding, in the past raw eggs have often been claimed to be an ideal source of calories, for reasons that sound logical. "An egg should never be cooked," wrote raw-foodists Molly and Eugene Christian in 1904. "In its natural state it is easily dissolved and readily taken up by all the organs of digestion, but the cooked egg must be brought back to liquid form before it can be digested, which puts extra and unnecessary labor upon those over-worked organs." This kind of argument persuaded generations of bodybuilders. The first muscleman with wide popular appeal was Steve Reeves, Hollywood's movie Hercules of the 1950s, who famously ate raw eggs every day for breakfast. Celebrated strongmen like Charles Atlas and Arnold Schwarzenegger touted their merits too—as Mr. Universe, Schwarzenegger swallowed his eggs mixed with thick cream. Raw egg–eating by muscular athletes has even entered popular culture. In 1976 Sylvester Stallone's boxing hero Rocky Balboa swallowed them as part of his training regimen in the movie *Rocky*. Thirty years later, in *Rocky Balboa*, he was still downing raw eggs. The quantity eaten by these legendary figures could be daunting: "Iron Guru" Vince Gironda, a popular teacher of bodybuilders, recommended up to thirty-six raw eggs a day.

Raw eggs would seem to provide an excellent food supply not only because their protein needs no chewing but also because their chemical composition is ideal. The amino acids of chicken eggs come in about forty proteins in almost exactly the proportions humans require. The match gives eggs a higher biological value—a measure of the rate at which the protein in food supports growth—than the protein of any other known food, even milk, meat, or soybeans. Raw eggs have other natural advantages. Their shells make them safer from bacterial contamination than cuts of meat. When aborigines on the beaches of Australia's tropical north coast are thirsty, they look for turtle nests and readily drink raw egg whites. Eggs are the only unprocessed animal food that can safely be stored at room temperature for several weeks.

But even though eggs appear to be both high-quality and relatively safe when eaten raw, hunter-gatherers prefer to cook them. Unlike Australians, the Yahgan hunter-gatherers of Tierra del Fuego "would never eat half-cooked, much less raw eggs." The Yahgan bored holes in eggshells to prevent them from bursting, buried the eggs on the edge of the fire, and turned them until they were quite hard inside. When not drinking eggs to slake their thirst, Australian aborigines would take similar pains, throwing emu eggs in the air to scramble them while still intact. They would then put them into hot sand or ashes and turn them regularly to cook them evenly, taking about twenty minutes. Such care suggests that the hunter-gatherers knew better than the musclemen.

In the late 1990s a Belgian team of gastroenterologists tested the effects of cooking for the first time, using a new research tool that allowed the investigators to follow the fate of egg proteins after they had been swallowed. The researchers fed hens a diet rich in stable isotopes of carbon, nitrogen, and hydrogen. The labeled atoms found their way into the eggs, allowing the experimenters to monitor the fate of protein molecules when the eggs were eaten. To determine how much of an egg meal was digested and absorbed in the body, they adopted the same method that had been used for studies of starch digestibility: they collected the food remains from the end of people's small intestine, the ileum. Any protein that was undigested by the time it reached the ileum was metabolically useless to the person who ate it, because in the large intestine bacteria and protozoa digest the food proteins entirely for their own benefit.

At first the experimenters worked only with ileostomy patients, but later they were able to check their results with healthy subjects as well. The ileostomy patients and healthy volunteers each ate about four raw or cooked eggs, containing a total of 25 grams (0.9 ounces) of protein. Results were similar for the two groups. When the eggs were cooked, the proportion of protein digested averaged 91 percent to 94 percent. This high figure was much as expected given that egg protein is known to be an excellent food. However, in the ileostomy patients, digestibility of raw eggs was measured at a meager 51 percent. It was a little higher, 65 percent, in the healthy volunteers whose protein digestion was estimated by the appearance of stable isotopes in the breath. The results showed that 35 percent to 49 percent of the ingested protein was leaving the small intestine undigested. Cooking increased the protein value of eggs by around 40 percent.

The Belgian scientists considered the reason for this dramatic effect on nutritional value and concluded that the major factor was denaturation of the food proteins, induced by heat. Denaturation occurs when the internal bonds of a protein weaken, causing the molecule to open up. As a result, the protein molecule loses its original three-dimensional structure and therefore its natural biological function. The gastroenterologists noted that heat predictably denatures proteins, and that denatured proteins are more digestible because their open structure exposes them to the action of digestive enzymes.

Even before the Belgian egg study, there were indications that cooking can be responsible for enough denaturation to strongly influence digestibility. In 1987 researchers chose to study a beef protein, bovine serum albumin (BSA), selected because it is a typical food protein. In cooked samples, digestion by the enzyme trypsin increased four times compared to that of uncooked samples. The researchers concluded that the simple process of denaturation by heat (causing the protein molecule to unfold and lose its solubility in water) explained its greatly increased susceptibility to digestion.

Heat is only one of several factors that promote denaturation. Three others are acidity, sodium chloride, and drying, all of which humans use in different ways.

Acid is vital in the ordinary process of digestion. Our empty stomachs are highly acidic thanks to the secretions of a billion acid-producing cells that line the stomach wall and secrete one to two liters of hydrochloric acid a day. Food entering the stomach buffers the acidity and causes a more neutral pH, but the stomach cells respond rapidly and secrete enough acid to return the stomach to its original intense low pH, less than 2. This intense acidity has at least three functions: it kills bacteria that enter with the food, activates the digestive enzyme pepsin, and denatures proteins. Denaturation looks particularly important.

Marinades, pickles, and lemon juice are acidic, so if applied for sufficient time they can contribute to the denaturing of proteins in meat, poultry, and fish. It is no surprise that we like seviche, raw fish marinated in a citrus juice mixture, traditionally for a few hours. Hunter-gatherers have likewise been reported mixing acidic fruits with stored meats. The Tlingit of Alaska stuffed goat meat with blueberries and stored salmon spawn mashed with cooked huckleberries. Many other North American groups made pemmican by mixing dried and pounded meat with various kinds of berries, and Australian aborigines mixed wild plums with the pounded bones and meat of kangaroo. While pleasing flavors and improved storage might be enough to account for such mixtures, increased digestibility could also contribute to explaining the broad use of these acidic preparations. Animal protein that has been salted and dried, such as fish, is likewise denatured and thereby made more digestible. Increased digestibility from denaturation also helps account for our enjoyment of dried meats such as jerky or salted fish.

Although gelatinization and denaturation are largely chemical effects, cooking also has physical effects on the energy food provides. Research on the topic began with a misfortune almost two hundred years ago. On June 6, 1822, twenty-eight-year-old Alexis St. Martin was accidentally shot from a distance of about a meter (three feet) inside a store of the American Fur Company at Fort Mackinac, Michigan. William Beaumont, a young, war-hardened surgeon, was nearby and arrived within twenty-five minutes to find a bloody scene that he described eleven years later: "A large portion of the side was blown off, the ribs fractured, and openings made into the cavities of the chest and abdomen, through which protruded portions of the lung and stomach, much lacerated and burnt, exhibiting altogether an appalling and hopeless case. The diaphragm was lacerated and perforation made directly into the cavity of the stomach, through which food was escaping at the time your memorialist was called to his relief."

Beaumont took St. Martin to his own home. To everyone's surprise, St. Martin survived, and Beaumont continued to house and care for him after he stabilized. In a few months the patient resumed a vigorous life, and he became so strong that he eventually even paddled his family in an open canoe from Mississippi to Montreal. Although the fist-sized wound mostly filled in, it never completely closed. For the rest of St. Martin's life, the inner workings of his stomach were visible from the outside.

The ambitious Beaumont realized that he had an extraordinary study opportunity. He began on August 1, 1825. "At 12 o'clock, M., I introduced through the perforation, into the stomach, the following articles of diet, suspended by a silk string, and fastened at proper distances, so as to pass in without pain—viz.:—a piece of highly seasoned *a la mode* beef; a piece of *raw, salted, fat pork*; a piece of *raw, salted, lean beef*; a piece of *boiled, salted beef*; a piece of *stale bread*; and a bunch of *raw, sliced cabbage*; each piece weighing about two drachms; the lad [St. Martin] continuing his usual employment around the house."

Beaumont observed the stomach closely. He noted how quiet it was when it had no food, the rugae (muscle folds) nestled upon each other. When soup was swallowed, the stomach was at first slow to respond. "The rugae gently close upon it, and gradually diffuse it through the gastric cavity." When Beaumont placed food directly on the stomach wall, the stomach became excited and its color brightened. There was a "gradual appearance of innumerable, very fine, lucid specks, rising through the transparent mucous coat, and seeming to burst, and discharge themselves upon the very points of the papillae, diffusing a limpid, thin fluid over the whole interior gastric surface." For the first time, it was possible to watch digestion in action.

Beaumont continued his experiments intermittently for eight years. He recorded in detail how long it took foods to be digested by the stomach and emptied into the duodenum.

From those observations he drew two conclusions relevant to the effects of cooking.

The more tender the food, the more rapidly and completely it was digested. He noted the same effect for food that was finely divided. "Vegetable, like animal substances, are more capable of digestion in proportion to the minuteness of their division . . . provided they are of a soft solid." Potatoes boiled to reduce them to a dry powder tasted poor, but they were more easily digested. If not powdered, entire pieces remained long undissolved in the stomach and yielded slowly to the action of the gastric juice. "The difference is quite obvious on submitting parcels of this vegetation, in different states of preparation, to the operation of the gastric juice, either in the stomach or out of it."

The same principles held, said Beaumont, with respect to meat. "Fibrine and gelatine [muscle fibers and collagen in meat] are affected in the same way. If tender and finely divided, they are disposed of readily; if in large and solid masses, digestion is proportionally retarded. . . . Minuteness of division and tenderness of fibre are the two grand essentials for speedy and easy digestion."

In addition to "minuteness of division and tenderness," cooking helped. He was explicit in the case of potatoes. "Pieces of raw potato, when submitted to the operation of this fluid, in the same manner, almost entirely resist its action. Many hours elapse before the slightest appearance of digestion is observable, and this only upon the surface, where the external laminae become a little softened, mucilaginous, and slightly farinaceous. Every physician who has had much practice in the diseases of children knows that partially boiled potatoes, when not sufficiently masticated (which is always the case with children), are frequently a source of colics and bowel complaints, and that large pieces of this vegetable pass the bowels untouched by digestion." It was the same with meat. When Beaumont introduced boiled beef and raw beef at noon, the boiled beef was gone by 2 P.M. But the piece of raw, salted, lean beef of the same size was only slightly macerated on the surface, while its general texture remained firm and intact.

Sadly, St. Martin came to resent being a focus of scientific interest. By the time of his death in 1880 at the ripe old age of eighty-five, he felt thoroughly mistreated. He had long refused to have anything to do with Beaumont, and his family shared his sense of abuse. Dr. William Osler, often described as the father of modern medicine, hoped to study St. Martin's body and even buy his stomach, but the family refused. They kept his body privately for four days to ensure that it rotted, then they buried him in an unusually deep grave, eight feet down, to thwart any medical interest in his organs.

Beaumont's discovery that soft and finely divided foods are more easily digested conforms to our preference for such items. In 2006 the London department store Selfridges received five advance orders for a new product: the world's most expensive sandwich. For £85 ($148) people had the chance to eat a 595-gram (21-ounce) mixture of fermented sourdough bread, Wagyu beef, fresh lobe foie gras, black truffle mayonnaise, brie de Meaux, English plum tomatoes, and confit. The beef explains the high price. Wagyu cattle are one of the most expensive breeds in the world because their meat is exceptionally tender, and no effort is spared to make it so. The animals are raised on a diet that includes beer and grain, and their muscles are regularly massaged with sake, the Japanese rice wine. The fat in the meat is claimed to melt at room temperature. The exceptional value of Wagyu beef illustrates a notable human pattern: people like their meat tender. "Of all the attributes of eating quality," wrote meat scientist R. A. Lawrie, "texture and tenderness are presently rated most important by the average consumer, and appear to be sought at the expense of flavour and colour." A key aim of meat science is to discover how to produce the most tender meat. Rearing, slaughtering, preservation, and preparation methods all play their part.

So does cooking. According to cooking historian Michael Symons, the cook's main goal has always been to soften food. "The central theme is that cooks assist the bodily machine," he wrote. He cited *Mrs. Beeton's Book of Household Management*, which in 1861 sought to advise naive housewives about the fundamentals of the kitchen. The first of six reasons for cooking was "to render mastication easy." "Hurrying over our meals, as we do, we should fare badly if all the grinding and subdividing of human food had to be accomplished by human teeth." A second reason for cooking stressed the point Beaumont had discovered: "to facilitate and hasten digestion."

The way Kalahari San hunter-gatherers prepare their food suggests a similar concern for making their meals as soft as possible. They cook their meat until "it is so tender that the sinews will fall apart." Then "it is usually crushed in a mortar." It is the same with plant foods. After melons or seeds have been cooked by burying them in hot embers or ashes, their contents are "ground in a mortar and eaten as a gruel."

Tropical and subtropical hunter-gatherers, such as Andaman Islanders, Siriono, Mbuti, and Kalahari San, eat all their meat cooked. It is in cooler climates that people sometimes eat animal protein raw. If they are eaten uncooked, the raw items tend to be soft, like the mammal livers and rotten fish the Inuit eat. The island-living Yahgan in the south of Tierra del Fuego have three such foods, according to Martin Gusinde, who lived with them for twenty years. There is "the soft meat" of mollusks such as winkles, "squeezed out of the calcareous shell with a slight pressure of the fingers and eaten without any preparation, except that occasionally the little morsel of fish is dipped into seal blubber." There are also the ovaries of sea urchins and the milky liquid in the shell, a delicacy shared by the Tlingit and eaten by Japanese and Europeans today in fine restaurants. According to Gusinde, a few individuals found the raw fat of a young whale tasty. Other than these cases, all animal protein was cooked.

Game animals have a few soft parts. The Utes of Colorado were said to roast all their meat but they ate the kidneys and livers raw. Australian aborigines supposedly eat mammal intestines raw on occasion, as Inuit do with fish and birds. Raw intestines may seem a startling preference in view of the potential for parasites to be present. They are likewise almost always the first part of a prey animal eaten by chimpanzees, chewed and swallowed much faster than muscle meat.

Raw-blood meals are well known among pastoralists such as Maasai, and as we saw in chapter 1, reported by Marco Polo in thirteenth-century Mongol nomad warriors. Elsewhere raw-fat meals are provided by fat-tailed sheep. Asian nomads value these sheep so highly and have bred them to such an extreme that they sometimes provide their animals with little carts to support the massive tail. On trek the nomads remove some of the fat for a raw meal, and the sheep travels a little lighter the next day.

While some foods are naturally tender, meat is variable. Meat with smaller muscle fibers is more tender, so chicken is more tender than beef. An animal slaughtered without being stressed retains more glycogen in its muscles. After death the glycogen converts to lactic acid, which promotes denaturation and therefore a more tender meat. Carcasses that are left to hang for several days are more tender, because proteins are partly broken down by enzymes.

But nothing changes meat tenderness as much as cooking because heat has a tremendous effect on the material in meat most responsible for its toughness: connective tissue.

Composed of a fibrous protein called collagen and a stretchy one called elastin, connective tissue wraps the meat in three pervasive layers. The innermost layer is a sleeve called endomysium, which surrounds each individual muscle fiber like the skin of a sausage. Bundles of endomysium-enclosed muscle fibers lie alongside one another jointly sheathed in a larger skin, the perimysium. Finally, those bundles, or fascicles, are held together by the outer wrapping, or epimysium, which encloses the entire muscle. At the end of the muscle, the epimysium turns into the tendon. Connective tissue is slippery, elastic, and strong: the tensile strength of tendons can be half that of aluminum. So connective tissue not only does a wonderful job of keeping our muscles in place but it also makes meat very difficult to eat, particularly for an animal like humans or chimpanzees whose teeth are notably blunt.

The main protein in connective tissue, collagen, owes its toughness to an elegant repeating structure. Three left-handed helices of protein twirl around one another to form a right-handed superhelix. The superhelixes join into fibrils, and the fibrils form fibers that assemble into a crisscross pattern. The effect is a marvel of microengineering. The extraordinary mechanical strength of collagen explains why sinews, or tendons, make excellent bowstrings and why it is the most abundant protein in vertebrates: it is the main component of skin.

But collagen has an Achilles' heel: heat turns it to jelly. Collagen shrinks when it reaches its denaturation temperature of 60–70°C (140–158°F), and then, as the helices start to unwind, it starts melting away. Whether heated about 100°C (212°F) for a short time or at lower temperatures for a longer time, the fibrils of collagen fall apart until they convert into the very antithesis of toughness: gelatin, a protein with commercial uses from Jell-O to jellied eels. The amount of force required to cut through a standard piece of meat tends to reach a minimum between 60°C and 70°C (140 and 158°F). Above those temperatures, slow cooking in water can sometimes continue to increase the tenderness.

Unfortunately for the amateur cooks among us, a second effect of heating meat is contrary to the first. Unlike connective tissue, heated muscle fibers tend to get tougher and drier. The cumulative effects of cooking meat are therefore complex. Bad cooking can render meat hard to chew, but good cooking tenderizes every kind of meat, from shrimp and octopus to rabbit, goat, and beef. Tenderness is even important for cooks preparing raw meat. Steak tartare requires a particularly high grade of meat (low in connective tissue) and the addition of raw eggs, onions, and sauces. The *Joy of Cooking* recommends grinding top sirloin, or scraping it with the back of a knife, until only the fibers of connective tissue remain.

Steak tartare supposedly gets its name from the Tartars, or Mongols, who rode in Genghis Khan's army. When soldiers were moving too fast to cook, they sometimes drank horse blood but they were also reported to put slabs of meat under the saddles, riding on them all day until they were tender.

Brillat-Savarin recorded an enthusiastic testimony of the practice: "Dining with a captain of Croats in 1815, 'Gads,' said he, 'there's no need of so much fuss in order to have a good dinner! When we are on scout duty and feel hungry, we shoot down the first beast that comes in our way, and cutting out a good thick slice, we sprinkle some salt over it, place it between the saddle and the horse's back, set off at a gallop for a sufficient time, and' (working his jaws like a man eating large mouthfuls) '*gniaw, gniaw, gniaw*, we have a dinner fit for a prince.'"

Why does tenderness matter? Beaumont observed that softer food was digested faster, and since faster or easier digestion demands less metabolic effort, softer food might lead to energy saved during digestion. The idea should make sense when you consider the greater liveliness you feel after eating a light meal compared to a heavy one: the light meal demands less work from your intestines and therefore makes other kinds of physical activity easy. This energy-saving principle has been beautifully shown in rats given soft food.

A team of Japanese scientists led by Kyoko Oka reared twenty rats on two different food regimes. Ten rats ate ordinary laboratory pellets, which were hard enough to require substantial chewing. The other ten ate a version of the standard food that was modified in a single way: the pellets were made softer by increasing their air content. The soft pellets were puffed up like a breakfast cereal and required only half the force of the hard pellets to crush them. In every other way the rats' conditions were identical. The calorie intake, and calorie expenditure on locomotion, were found to be the same for the two groups. The ordinary and soft pellets did not differ in how much they had been cooked, their nutrient composition, or water content. Conventional theory based on the calculation of calorie intake would predict that the two groups of rats should have grown at the same rates and to the same size. They should have had the same body weight and the same levels of fat.

But they did not. The rats began eating their different pellet diets at four weeks old. By fifteen weeks the growth curves of the two groups had visibly separated, and by twenty-two weeks the group curves were significantly different. The rats eating soft food slowly became heavier than those eating hard food: on average, 37 grams heavier, or about 6 percent; and they had more abdominal fat: on average, 30 percent more, enough to be classified as obese. Soft, well-processed foods made the rats fat. The difference was in the cost of digestion. At every meal the rats experienced a rise in body temperature, but the rise was lower in the soft-pellet group than in the hard-pellet group. The difference was particularly strong in the first hour after eating, when the stomach was actively churning and secreting. The researchers concluded that the reason the softer diet led to obesity was simply that it was a little less costly to digest.

The implications of Oka's experiment are clear. If cooking softens food and softer food leads to greater energy gain, then humans should get more energy from cooked food than raw food not only because of processes such as gelatinization and denaturation, but also because it reduces the costs of digestion. This prediction has been studied in the Burmese python. Physiological ecologist Stephen Secor finds pythons to be superb experimental subjects because after swallowing a meal, the snakes lie in a cage doing little but digesting and breathing. By measuring how much oxygen the pythons consume before and after a meal, Secor measures precisely how much energy the snakes use, and can attribute it to the cost of digestion. He typically monitors the snakes for at least two weeks at a time.

Secor and his team have shown repeatedly that the physical structure of a python's diet influences its cost of digestion. If the snake eats an intact rat, its metabolic rate increases more than if a similar rat is ground up before the snake eats it. Amphibians yield the same results. Toads given hard-bodied worms have higher costs of digestion than those eating soft-bodied worms. Just as Oka's team found with rats eating softer pellets, Secor's studies show that softer meat is also digested with less energy expenditure.

A particular advantage of the Burmese pythons is that experimenters can insert food directly into their esophagus. The snakes show no signs of objecting. No matter whether the pythons find a food appealing and regardless of how easy the food is to swallow, the pythons just digest what they are given. They are an ideal species in which to test the effects of cooking on the cost of digestion. I approached Secor in 2005 to ask if he would be interested in the following study. Secor assigned eight snakes to the research, and his team prepared five kinds of experimental diet. Lean beef steak (eye of round, with less than 5 percent fat) was the basic food and was given to the snakes in each of four preparations: raw and intact; raw and ground; cooked and intact; and cooked and ground. The snakes were also given whole intact rats.

The experiment took several months. As expected from earlier results, the snakes' cost of digestion when they ate the raw, intact meat was the same as for the whole rats. But grinding and cooking changed the costs of digestion. Grinding breaks up both muscle fibers and connective tissue, so it increases the surface area of the digestible parts of the meat. Ground meat is exposed more rapidly to acid, causing denaturation, as well as to proteolytic enzymes, causing degradation of the muscle proteins. Grinding reduced the snakes' cost of digestion by 12.3 percent. Cooking produced almost identical results. Compared to the raw diet, cooked meat led to a reduction in the cost of digestion by 12.7 percent. The effects of the two experimental treatments, grinding and cooking, were almost entirely independent. Alone, each reduced the cost of digestion by just over 12 percent. Together, they reduced it by 23.4 percent.

Mrs. Beeton was right to cherish softness as an aid to digestion. It makes sense that we like foods that have been softened by cooking, just as we like them chopped up in a blender, ground in a mill, or pounded in a mortar. The unnaturally, atypically soft foods that compose the human diet have given our species an energetic edge, sparing us much of the hard work of digestion. Fire does a job our bodies would otherwise have to do. Eat a properly cooked steak, and your stomach will more quickly return to quiescence. From starch gelatinization to protein denaturation and the costs of digesting, absorbing, and assimilating meat, the same lesson emerges. Cooking gives calories.

When we consider the difficulties humans experience on raw diets, the evidence that all animals thrive on cooked food, and the nutritional evidence concerning gelatinization, denaturation, and tenderness, what is extraordinary about this simple claim is that it is new. Admittedly, cooking can have some negative effects. It leads to energy losses through dripping during the cooking process and by producing indigestible protein compounds, and it often leads to a reduction of vitamins. But compared to the energetic gains, those processes do not matter. Overall it appears that cooking consistently provides more energy, whether from plant or animal food.

Why then do we like cooked food today? The energy it provides is more than many of us need, but it was a critical contribution for our remote ancestors just as it is vital for many people living nowadays in poverty. Tens of thousands of generations of eating cooked food have strengthened our love for it. Consider foie gras, the liver of French geese that have been cruelly force-fed to make them especially fat. The fresh liver is soaked in milk, water, or port, marinated in Armagnac, port, or Madeira, seasoned, and finally baked. The result is so meltingly soft and tender that a single bite has been said to make a grown man cry. Our raw-food-eating ancestors never knew such joy.

Cooked food is better than raw food because life is mostly concerned with energy. So from an evolutionary perspective, if cooking causes a loss of vitamins or creates a few long-term toxic compounds, the effect is relatively unimportant compared to the impact of more calories. A female chimpanzee with a better diet gives birth more often and her offspring have better survival rates. In subsistence cultures, better-fed mothers have more and healthier children. In addition to more offspring, they have greater competitive ability, better survival, and longer lives. When our ancestors first obtained extra calories by cooking their food, they and their descendants passed on more genes than others of their species who ate raw. The result was a new evolutionary opportunity.

CHAPTER 4

When Cooking Began

"The introduction of cooking may well have been the decisive factor in leading man from a primarily animal existence into one that was more fully human."

—Carleton S. Coon, *The History of Man*

Archaeologists are divided about the origins of cooking. Some suggest that fire was not regularly used for cooking until the Upper Paleolithic, about forty thousand years ago, a time when people were so modern that they were creating cave art. Others favor much earlier times, half a million years ago or before. A common proposal lies between those extremes, advocated especially by physical anthropologist Loring Brace, who has long noted that people definitely controlled fire by two hundred thousand years ago and argues that cooking started around the same time. As the wide range of views shows, the archaeological evidence is not definitive. Archaeology offers only one safe conclusion: it does not tell us what we want to know. But though we cannot solve the problem of when cooking began by relying on the faint traces of ancient fires, we can use biology instead. In the teeth and bones of our ancestors we find indirect evidence of changes in diet and the way it was processed.

Yet the archaeological data leave no doubt that controlling fire is an ancient tradition. In the most recent quarter of a million years, there is sparkling evidence of fire control, and even occasionally of cooking, by both our ancestors and our close relatives the Neanderthals. The most informative sites tend to be airy caves or rock shelters, many of them in Europe. In Abri Pataud in France's Dordogne region, heat-cracked river cobblestones from the late Aurignacian period, around forty thousand years ago, show that people boiled water by dropping hot rocks in it. At Abri Romani near Barcelona, a series of occupations dating back seventy-six thousand years includes more than sixty hearths together with abundant charcoal, burnt bones, and casts of wooden objects possibly used for cooking. More than ninety-three thousand years ago in Vanguard Cave, Gibraltar, three separate episodes of burning can be distinguished in a single hearth. Neanderthals heated pinecones on these fires and broke them open with stones, much as contemporary hunter-gatherers have been recorded doing, to eat the seeds.

Our ancestors were using fire in the Middle East and Africa as well. In a cave at Klasies River Mouth, a coastal site in South Africa from sixty thousand to ninety thousand years ago, burnt shells and fish bones lie near family-size hearths that appear to have been used for weeks or months at a time. Between 109,000 and 127,000 years ago in the Sodmein Cave of Egypt's Red Sea Mountains, people appear responsible for huge fires with three distinct superimposed ash layers and the burnt bones of an elephant. Charred logs, together with charcoal, reddened areas, and carbonized grass stems and plants, date to 180,000 years ago at Kalambo Falls in Zambia. Back to 250,000 years ago in Israel's Hayonim Cave, there are abundant hearths with ash deposits up to 4 centimeters (1.6 inches) thick. Such sites show that people have been controlling fire throughout the evolutionary life span of our species, *Homo sapiens*, which is considered to have originated about two hundred thousand years ago.

Because evidence about controlling fire is inconsistent before the last quarter of a million years, it is often argued that the control of fire was unimportant or absent until that time. But that idea is now particularly shaky because the older part of the record, going back in time from a quarter of a million years ago, has been improving in quality. Two sites in particular give tantalizing hints of what earlier people were doing with fire.

An ancient fireplace at Beeches Pit archaeological site in England securely dated to four hundred thousand years ago lies on the gently sloping bank of an ancient pond. Eight hand axes attest to the presence of humans. Dark patches about one meter (three feet) in diameter with reddened sediments at the margins show where burning occurred. Tails of ashlike material lead down from the fires toward the pond, while the upper side contains numerous pieces of flint. The flints have been knapped, or broken by a sharp blow, and many are burnt. A team led by archaeologist John Gowlett fitted the flint pieces together, and one of the various refits showed that someone had been knapping a heavy core (1.3 kilograms, or 2.9 pounds) until a flaw became obvious. The knapper abandoned it, and two flakes from the series fell forward and were burnt, indicating that the toolmaker apparently had been squatting next to a warming blaze.

Another four-hundred-thousand-year-old site, at Schöningen in Germany, has yielded more than a half dozen superb throwing spears carved from spruce and pine, together with the remains of at least twenty-two horses that appear to have died at the same time as one another, apparently killed by humans. Cut marks show that people removed meat from the horses. At the same site were numerous pieces of burnt flint, four large reddened patches about one meter in diameter that appear to have been fireplaces, and some pieces of burnt wood including a shaped stick, also made from spruce, that had been charred at one end as if it had been used as a poker, or perhaps held over coals to cook strips of meat. This exceptional lakeshore find by archaeologist Hartmut Thieme represents the earliest evidence of group hunting. Thieme suggests that after people killed the horse group, they found themselves with far more food than they could consume at the time. They settled for several days and built the fires along the lakeshore to dry as much meat as possible.

Prior to half a million years ago, there is no evidence for the control of fire in Europe, but ice covered Britain for much of the time between five hundred thousand and four hundred thousand years ago, and glaciers would have swept away most evidence of any earlier occupations. Farther south, however, fire-using is strongly attested at 790,000 years ago. In a well-dated site called Gesher Benot Ya'aqov, next to Israel's Jordan River, hand axes and bones were first discovered in the 1930s, and in the 1990s, Naama Goren-Inbar found burnt seeds, wood, and flint. Olives, barley, and grapes were among the species of seeds found burned. The flint fragments were grouped in clusters, suggesting they had fallen into campfires. Nira Alperson-Afil analyzed these dense concentrations. She concluded that the early humans who made these fires "had a profound knowledge of fire-making, enabling them to make fire at will."

Gesher Benot Ya'aqov is the oldest site offering confident evidence of fire control. Before then we find only provocative hints. Archaeological sites between a million and a million and a half years old include burnt bones (at Swartkrans in South Africa), lumps of clay heated to the high temperatures associated with campfires (Chesowanja, near Lake Baringo in Kenya), heated rocks in a hearthlike pattern (Gadeb in Ethiopia), or colored patches with appropriate plant phytoliths inside (Koobi Fora, Kenya). But the meaning of such evidence as indicating human control of fire is disputed. Some archaeologists find it totally unconvincing, regarding natural processes such as lightning strikes as likely explanations for the apparent use of fire. Others accept the idea that humans controlled fire in the early days of *Homo erectus* as well established. Overall, these hints from the Lower Paleolithic tell us only that in each case the control of fire was a possibility, not a certainty.

Evidence of humans controlling fire is hard to recover from early times. Meat can be cooked easily without burning bones. Fires might have been small, temporary affairs, leaving no trace within a few days of exposure to wind and rain. Even now hunter-gatherers such as the Hadza, who live near the Serengeti National Park in northern Tanzania, may use a fire only once, and they often leave no bones or tools at the fire site, so archaeologists would not be able to infer human activity even if they could detect where burning had occurred. The caves and shelters that preserve relatively recent evidence of fire use tend to be made of soft rock, such as limestone, which erodes quickly, so the half-lives of caves average about a quarter of a million years, leaving increasingly few opportunities to find traces of fire use from earlier periods. From the past quarter of a million years there are sites of human occupation where people must have used fire, yet there is no sign of it. There are also mysterious reductions in the frequency of finding evidence of fire, such as one that followed an interglacial period in Europe from 427,000 years ago to 364,000 years ago, when fire evidence was relatively abundant. In short, while humans have certainly been using fire for hundreds of thousands of years, archaeology does not tell us exactly when our ancestors began to do so.

The inability of the archaeological evidence to tell when humans first controlled fire directs us to biology, where we find two vital clues. First, the fossil record presents a reasonably clear picture of the changes in human anatomy over the past two million years. It tells us what were the major changes in our ancestors' anatomy, and when they happened. Second, in response to a major change in diet, species tend to exhibit rapid and obvious changes in their anatomy. Animals are superbly adapted to their diets, and over evolutionary time the tight fit between food and anatomy is driven by food rather than by the animal's characteristics. Fleas do not suck blood because they happen to have a proboscis well designed for piercing mammalian skin; they have the proboscis because they are adapted to sucking blood. Horses do not eat grass because they happen to have the right kind of teeth and guts for doing so; they have tall teeth and long guts because they are adapted to eating grass. Humans do not eat cooked food because we have the right kind of teeth and guts; rather, we have small teeth and short guts as a result of adapting to a cooked diet.

Therefore, we can identify when cooking began by searching the fossil record. At some time our ancestors' anatomy changed to accommodate a cooked diet. The change must mark when cooking became not merely an occasional activity but a predictable daily occurrence, because until then our ancestors would have sometimes had to resort to eating their food raw—and therefore could not adapt to cooking. The time when our ancestors became adapted to cooked food also marks the time when fire was controlled so effectively that it was never lost again.

Anthropologists have sometimes suggested that humans could have controlled fire for reasons such as warmth and light for many millennia before starting to use it for cooking. However, many animals show a spontaneous preference for cooked food over raw. Would prehuman ancestors have preferred cooked food also? Evolutionary anthropologists Victoria Wobber and Brian Hare tested chimpanzees and other apes in the United States, Germany, and Tchimpounga, a Congolese sanctuary. Across the different locations, despite different diets and living conditions, the apes responded similarly. No apes preferred any food raw. They ate sweet potatoes and apples with equal enthusiasm whether raw or cooked, but they preferred their carrots, potatoes, and meat to be cooked. The Tchimpounga chimpanzees were particularly informative because there was no record of them having eaten meat previously, yet they showed a strong preference for cooked meat over raw meat. The first of our ancestors to control fire would likely have reacted the same way. Cooked food would have suited their palate the first time they tried it, just as a taste for cooked food, with its immediate benefits, is shared by a wide range of wild and domestic species. Chimpanzees in Senegal do not eat the raw beans of *Afzelia* trees, but after a forest fire has passed through the savanna, they search under *Afzelia* trees and eat the cooked seeds.

Why are wild animals pre-adapted in this way to appreciate the smells, tastes, and textures of cooked food? The spontaneous preference for cooked food implies an innate mechanism for recognizing high-energy foods. Many foods change their taste when cooked, becoming sweeter, less bitter, or less astringent, so taste could play a role in this preference, as some evidence suggests. Koko is a gorilla who learned to communicate with humans, and she prefers her food cooked. Cognitive psychologist Penny Patterson asked her why: "I asked Koko while the video was rolling if she liked her vegetables better cooked (specifying my left hand) or raw/fresh (indicating my right hand). She touched my left hand (cooked) in reply. Then I asked why she liked vegetables better cooked, one hand standing for 'tastes better,' the other 'easier to eat.' Koko indicated the 'tastes better' option."

When primates eat, sensory nerves in the tongue perceive not only taste but also particle size and texture. Some of the brain cells (neurons) responsive to texture converge with taste neurons in the amygdala and orbito-frontal cortex of the brain, allowing a summed assessment of food properties. This sensory-neural system enables primates to respond instinctively to a wide range of food properties other than merely taste, including such factors as grittiness, viscosity, oiliness, and temperature.

In 2004 such abilities in the human brain were reported for the first time. A team led by psychologist Edmund Rolls found that when people had foods of a particular viscosity in their mouths, specific brain regions were activated. Those regions partly overlapped with regions of taste cortex that register sweetness. The picture emerging from such studies is that hard-wired responses to properties such as taste, texture, and temperature are integrated in the brain with learned responses to the sight and smell of food. So the mechanisms that allow animals to assess the quality of raw foods directly apply to cooked foods and allow them to choose foods of a good texture for easy digestion.

Rolls's studies suggest that the proximate reasons chimpanzees and many other species like their meat and potatoes cooked may be the same as in humans. We identify foods that have high caloric value not just by their being sweet, but also by their being soft and tender. Our ancestors were surely prepared by their preexisting sensory and brain mechanisms to like cooked foods in the same way. A long delay between the first control of fire and the first eating of cooked food is therefore deeply improbable.

A long delay between the adoption of a major new diet and resulting changes in anatomy is also unlikely. Studies of Galapagos finches by Peter and Rosemary Grant showed that during a year when finches experienced an intense food shortage caused by an extended drought, the birds that were best able to eat large and hard seeds—those birds with the largest beaks—survived best. The selection pressure against small-beaked birds was so intense that only 15 percent of birds survived and the species as a whole developed measurably larger beaks within a year. Correlations in beak size between parents and offspring showed that the changes were inherited. Beak size fell again after the food supply returned to normal, but it took about fifteen years for the genetic changes the drought had imposed to reverse.

The Grants' finches show that anatomy can evolve very quickly in response to dietary changes. In the case of the drought year in the Galapagos, the change in diet was temporary and therefore so was the change in anatomy. Other data show that if an ecological change is permanent, the species also changes permanently, and again the transition is fast. Some of the clearest examples come from animals confined on islands that have been newly created by a rise in sea level. In fewer than eight thousand years, mainland boa constrictors that occupied new islands off Belize shifted their diets away from mammals and toward birds, spent more time in trees, became more slender, lost a previous size difference between females and males, and were reduced to a fifth of their original body weight. According to evolutionary biologist Stephen Jay Gould, this rate of change is not unusual. Drawing from the fossil record, he suggested that fifteen thousand to twenty thousand years may be about the average time one species takes to make a complete evolutionary transition to another. While a species that takes many years to mature, such as our ancestors, would take longer to evolve than a rapidly growing species, such rapid rates of evolution are sharply inconsistent with some previous interpretations of the effects of cooking. Loring Brace suggested that the use of fire for softening meat began around 250,000 to 300,000 years ago, followed by a supposed drop in tooth size that began about 100,000 years ago. This would mean that for at least the first 150,000 years after cooking was adopted, human teeth showed no response. Because such a long delay before adapting to a major new influence does not fit the animal pattern, we can conclude that Brace's idea is wrong. The adaptive changes brought on by the adoption of cooking would surely have been rapid.

In addition to following quickly, the changes would have been substantial. We can infer this from pairs of species in which lesser differences in diet have large effects. Take chimpanzees and gorillas, two ape species that often share the same forest habitat. In many ways their diets are very similar. Both choose ripe fruits when they are available. Both also supplement their diets with fibrous foods, such as piths and leaves. There is only one important difference in their food choice. When fruits are scarce, gorillas rely on foliage alone, whereas chimpanzees continue to search for fruit every day. Unlike gorillas, chimpanzees never survive only on piths and leaves—presumably because they are physiologically unable to do so.

The relative ability of these two apes to rely on foliage might at first glance appear to be a trivial matter—especially compared to the introduction of cooking. But many consequences follow from it. To find their vital fruits, chimpanzees must travel farther than gorillas, so they are more agile and smaller. There are differences in distributional range. Unlike chimpanzees, gorillas successfully occupy high-altitude forests without fruits, such as the Virunga Volcanoes of Rwanda, Uganda, and the Democratic Republic of Congo. Chimpanzees are limited to lower altitudes. Like other primates that are able to rely on a leaf diet, gorillas mature earlier, start having babies at a younger age, and reproduce faster.

Grouping patterns of these species also differ strikingly as a result of the difference in diet. The terrestrial foliage gorillas rely on is easily found and occurs in big patches, allowing their groups to be stable all year. But during food-poor seasons, chimpanzees are driven to travel alone or in small groups as they search for rare fruits. The difference in grouping patterns has further consequences. Gorillas form long-lasting bonds between females and males, whereas chimpanzees do not.

More than the relatively slight dietary difference that distinguishes gorillas from chimpanzees, cooked food had multiple differences from raw food. Effects of cooking include extra energy, softer food, fireside meals, a safer and more diverse set of food species, and a more predictable food supply during periods of scarcity. Cooking would therefore be expected to increase survival, especially of the vulnerable young. It should also have increased the range of edible foods, allowing extension into new biogeographical zones. The anatomical differences between a cooking and a pre-cooking ancestor should be at least as big as those between a chimpanzee and a gorilla. So whenever cooking was adopted, its effects should be easy to find. We can expect the origin of cooking to be signaled by large, rapid changes in human anatomy appropriate to a softer and more energy-rich diet.

The search for such changes proves to be rather simple. Before two million years ago, there is no suggestion for the control of fire. Since then there have been only three periods when our ancestors' evolution was fast and strong enough to justify changes in the species names. They are the times that produced *Homo erectus* (1.8 million years ago), *Homo heidelbergensis* (800,000 years ago), and *Homo sapiens* (200,000 years ago). These are therefore the only times when it is reasonable to infer that cooking could have been adopted.

Most recent was the evolution of *Homo sapiens* from an ancestor that is now usually called *Homo heidelbergensis.* It was a gentle process that began in Africa as early as three hundred thousand years ago and was largely complete by around two hundred thousand years ago. The transition was too recent to correspond to the origin of cooking, however, because *Homo heidelbergensis* was already using fire at Beeches Pit, Schöningen, and elsewhere four hundred thousand years ago. Nor does the transition to *Homo sapiens* show the kinds of change we are looking for. *Homo heidelbergensis* was merely a more robust form of human than *Homo sapiens,* with a large face, less rounded head, and slightly smaller brain. Most of the differences between these two species are too small and not obviously related to diet. We can be confident that cooking began more than three hundred thousand years ago, before *Homo sapiens* emerged.

Homo heidelbergensis evolved from *Homo erectus* in Africa from eight to six hundred thousand years ago. The timing of the *erectus–heidelbergensis* transition provides a reasonably comfortable fit with the archaeological data on the control of fire becoming particularly scarce. The main changes in anatomy from *Homo erectus* to *Homo heidelbergensis* were an increase in cranial capacity (brain volume) of around 30 percent, a higher forehead, and a flatter face. These are smaller modifications than the differences between a chimpanzee and a gorilla, and the modifications show little correspondence to changes in the diet. So this Pleistocene transition does not look favorable. It is a possibility for when cooking began, but not a promising one.

The only other option is the original change, from habilines to *Homo erectus*. This shift happened between 1.9 million and 1.8 million years ago and involved much larger changes in anatomy than any subsequent transitions. Recall that in many ways habilines were apelike. Like the australopithecines, they appear to have had two effective styles of locomotion. They walked upright and can be reconstructed as having had sufficiently strong and mobile arms to be good climbers. Their small size must have helped them in trees. They are estimated to have stood about 1 to 1.3 meters tall (3 feet 3 inches to 4 feet 3 inches) and appear to have weighed about the same as a chimpanzee, around thirty-two kilograms (seventy pounds) for a female and thirty-seven kilograms (eighty-one pounds) for a male. Despite their small bodies, they had much bigger chewing teeth than in any subsequent species of the genus *Homo*: the surface areas of three representative chewing teeth decreased by 21 percent from habilines to early *Homo erectus*. Habilines' larger teeth imply a bulky diet that required a lot of chewing.

Homo erectus did not exhibit the apelike features of the habilines. In the evolution of *Homo erectus* from habilines, we find the largest reduction in tooth size in the last six million years of human evolution, the largest increase in body size, and a disappearance of the shoulder, arm, and trunk adaptations that apparently enabled habilines to climb well. Additionally, *Homo erectus* had a less flared rib cage and a narrower pelvis than the australopithecines, both features indicating that they had a smaller gut. There was a 42 percent increase in cranial capacity. *Homo erectus* was also the first species in our lineage to extend its range beyond Africa: it was recorded in western Asia by 1.7 million years ago, Indonesia in Southeast Asia by 1.6 million years ago, and Spain by 1.4 million years ago. The reduction in tooth size, the signs of increased energy availability in larger brains and bodies, the indication of smaller guts, and the ability to exploit new kinds of habitat all support the idea that cooking was responsible for the evolution of *Homo erectus*.

Even the reduction in climbing ability fits the hypothesis that *Homo erectus* cooked. *Homo erectus* presumably climbed no better than modern humans do, unlike the agile habilines. This shift suggests that *Homo erectus* slept on the ground, a novel behavior that would have depended on their controlling fire to provide light to see predators and scare them away. Primates hardly ever sleep on the ground. Smaller species sleep in tree holes, in hidden nests, on branches hanging over water, on cliff ledges, or in trees so tall that no ground predator is likely to reach them. Great apes mostly build sleeping platforms or nests. The only non-human primate that regularly sleeps on the ground is the largest species of great ape, gorillas. Gorillas are safer on the ground than *Homo erectus* would have been because gorillas live in forests with few predators and they are relatively enormous. The most frequent ground sleepers are adult males, weighing around 127 kilograms (286 pounds). Smaller gorillas often sleep in trees.

The late Pliocene and early Pleistocene periods in Africa were rich in predators. In wooded areas from 4 million to 1.5 million years ago, our ancestors would have found saber-toothed cats. There was *Megantereon*, the size of a leopard, and *Dinofelis*, as big as a lion. In more open habitats there was the scimitar cat *Homotherium*, equally large. An extinct kind of lion and spotted hyena lived alongside our early ancestors, while modern lions and leopards have been present since at least 1.8 million years ago. There were also many large animals such as elephants, rhinoceroses, and buffalo-like ungulates that could stumble unawares onto an unconscious biped. The African woodlands would have been a very dangerous place to sleep on the ground.

Extrapolating from the behavior of living primates in predator-rich environments, the australopithecines and habilines surely slept in trees. Their habitats were well wooded and their upper-body anatomy suggests they climbed well. But what did *Homo erectus* do? The famous "Turkana boy," a beautifully preserved specimen of *Homo erectus* dated between 1.51 and 1.56 million years ago provides excellent evidence that they climbed relatively poorly. Physical anthropologists Alan Walker and Pat Shipman have described the Turkana boy as committed to locomotion on the ground. His finger bones had lost the curved, robust shape of australopithecine fingers. His shoulder blade had the modern form, giving no indication of being adapted to the stresses of climbing with the arm above the shoulder. The Turkana boy is so well preserved that Walker was able to study the vestibular system of the inner ear, responsible for balance. Species that climb regularly have a large and characteristically shaped vestibular system. The Turkana boy's is different from that of species that climb, but closely resembles the modern human system.

So the Turkana boy, like other *Homo erectus*, could not have climbed well and he therefore would have found it difficult to make the type of nest great apes sleep in. Chimpanzees take about five minutes to build their nests by standing on all fours where the nest is taking shape, bending branches toward themselves. They break some of the bigger ones and weave the branches together to form a platform that they finish off with a few leafy twigs that serve as cushions or pillows to make it comfortable. Making a nest depends on being able to move around easily on the end of a swaying branch. The long legs and flat feet of humans such as *Homo erectus* and modern people do not allow such agility. For a mother with a small infant, the gymnastic challenges of making a nest would have been particularly difficult given her need to cradle while she swayed in the tree.

Homo erectus therefore must have slept on the ground. But to do so in the dark of a moonless night seems impossibly dangerous. *Homo erectus* was as poorly defended a creature as we are, unable to sprint fast and dependent on weapons for any success in fighting. Surprised by a *Dinofelis* or a pack of hyenas at midnight, they would have been vulnerable.

If *Homo erectus* used fire, however, they could sleep in the same way as people do nowadays in the savanna. In the bush, people lie close to the fire and for most or all of the night someone is awake. When a sleeper awakens, he or she might poke at the fire and chat a while, allowing another to fall asleep. In a twelve-hour night with no light other than what the fire provides, there is no need to have a continuous eight-hour sleep. An informal system of guarding easily emerges that allows enough hours of sleep for all while ensuring the presence of an alert sentinel. To judge from records of attacks by jaguars, modern hunter-gatherers are safer in camp at night than they are on the hunt by day.

The control of fire could explain why *Homo erectus* lost their climbing ability. The normal assumption is that when long legs were favored, perhaps as a result of the increasing importance of long-distance travel as humans searched for meat, it was harder for humans to climb efficiently, and *Homo erectus* therefore abandoned the trees. But since that argument does not explain how *Homo erectus* could sleep safely, I prefer an alternative hypothesis: having controlled fire, a group of habilines learned that they could sleep safely on the ground. Their new practice of cooking roots and meat meant that food obtained from trees was less important than it had been when raw food was the only option. When they no longer needed to climb trees to find food or sleep safely, natural selection rapidly favored the anatomical changes that facilitated long-distance locomotion and led to living completely on the ground.

Two kinds of evidence thus point independently to the origin of *Homo erectus* as the time when cooking began. First, anatomical changes related to diet, including the reduction in tooth size and in the flaring of the rib cage, were larger than at any other time in human evolution, and they fit the theory that the nutritional quality of the diet improved and the food consumed was softer. Second, the loss of traits allowing efficient climbing marked a commitment to sleeping on the ground that is hard to explain without the control of fire.

The only alternative is the traditional theory that cooking was first practiced by beings that already looked like us—physically human members of the genus *Homo*. If this were true, by the time our ancestors adopted cooking, *Homo erectus* had long ago adapted to a soft, easily chewed diet of high caloric density. But as we have seen, cold-processing techniques such as grinding and blending provide relatively poor energy even when carried out by raw-foodists with modern equipment.

For more than 2.5 million years our ancestors have been cutting meat off animal bones, and the impact was huge. A diet that included raw meat as well as plant foods pushed our forebears out of the australopithecine rut, initiated the evolution of their larger brains, and probably inspired a series of food-processing innovations. But according to the evidence carried in our bodies, it would take the invention of cooking to convert habilines into *Homo erectus*, and launch the journey that has led without any major changes to the anatomy of modern humans.

CHAPTER 5

Brain Foods

"Tell me what you eat, and I shall tell you what you are."

—Jean Anthelme Brillat-Savarin, *The Physiology of Taste: Or Meditations on Transcendental Gastronomy*

"Man is but a reed, the weakest in nature, but he is a thinking reed," wrote philosopher Blaise Pascal in 1670. Exceptional intelligence is the defining feature of our species, yet its origins have long been a puzzle. Darwin concluded that intellect would have given advantages in social competition and the struggle to survive, but why humans should be brainier than other species was unclear. Only recently has an explanation emerged. In the view of many evolutionary anthropologists, the pressure for intelligence indeed comes primarily from the advantages of outwitting social competitors, whereas a major reason for species differences is how much brainpower the body can afford. For this reason the quality of the diet has been identified as a key driver of the growth of primate brains. For humans, cooking must have played a major role.

Attempts to explain the evolution of intelligence have sometimes appealed to rather specific advantages. Evolutionary biologist Richard Alexander argues that because humans practice warfare, and brainpower is critical for planning raids and winning battles, higher intellect could have been favored by a long evolutionary history of intense intergroup violence. But this hypothesis is undermined by chimpanzees, which behave in ways similar to warfare in small-scale human societies, but without humans' braininess. Violence between groups of chimpanzees is like a "shoot-on-sight" policy. Parties of males attack vulnerable rivals from adjacent groups whenever they encounter them, sometimes during incursions deep into the other group's territory in search of victims. Death rates from these interactions among chimpanzees are similar to those in small-scale societies of humans, yet chimpanzees are much less brainy than humans, and only about as clever as their more peaceable relatives, bonobos, gorillas, and orangutans.

Another suggested explanation for the evolution of intelligence is more ecological than social. This line of thinking proposes that intellect would be favored in species that occupy large home ranges, on the theory that wide-roaming creatures would need exceptional brainpower to mentally map their territories. And indeed, human hunter-gatherers cover huge areas compared to the ranges of apes and monkeys. But the correlation between range size and brain size does not generalize. Species of primates with larger brains are more intelligent, but they show no overall tendency to have larger ranges. The association of intellect and range size in humans looks accidental; that is, there is no evidence for a causal effect of brain size on range size, or vice versa, across primate species.

A more promising approach assumes that numerous kinds of benefits come from being intelligent. Clever species can forage in a variety of creative ways, such as using grasses and twigs to extract insects from holes, or lifting stones as hammers to smash nuts. Big-brained species can also manage complex social relationships. Evolutionary psychologist Robin Dunbar found that primates with bigger brains or more neocortex live in larger groups, have a greater number of close social relationships, and use coalitions more effectively than those with smaller brains.

Brains pay off socially when they beat brawn. Relationships can change daily in primates that live in large groups, such as chimpanzees or baboons. Flexible coalitions in which two or more group members gang up on another group member allow small or individually low-status animals to compete successfully for access to resources and mates. Coalitions are difficult to manage because individuals compete for the best allies, and an ally today may be a rival tomorrow. Individuals must constantly reassess one another's moods and strategies, and alter their own behavior accordingly. Clever animals can be deceitful too, deliberately hiding their feelings by masking facial expressions, or screaming to pretend they have been attacked when their real motive is to rally supporters to chase a dominant individual away from food. The result is a soap opera of changing affections, alliances, and hostilities, and a constant pressure to outsmart others.

Most animals are not up to the cognitive challenges of juggling social alliances. They compete one-on-one, like chickens, or following simple rules such as supporting members of their own group against outsiders. The exceptions are telling. Birds in the crow family have many of the social abilities of primates and are distinctly large-brained compared to other birds. Bottlenose dolphins form particularly complex and changeable alliances, and have the largest brains relative to body size of any nonhuman. Spotted hyenas live in large groups and use flexible coalitions to compete for power, and consistent with the primate evidence, they have bigger brains than their less social relatives. A similar link of sociality to mental power is found in social insects, whose neural tissue is concentrated not in brains but in ganglia. Darwin noted that colony-living ants and wasps have "cerebral ganglia of extraordinary dimensions," many times larger than other insects.

These kinds of correlation have supported the social brain hypothesis, which says that large brains have evolved because intelligence is a vital component of social life. The hypothesis nicely explains how animals that live in groups can benefit from being clever by outwitting their rivals in competition over mates, food, allies, and status. It also explains why species with bigger brains tend to have more complex societies, and the hypothesis suggests that if a species has limited brainpower, its social options may be constrained as well: small-brained monkeys may be too dim to handle many social relationships.

The social brain hypothesis is very important in explaining a major benefit of being intelligent. Indeed, the advantages are so strong that we might expect all social primates to have developed big brains and high intellect. Yet there is wide variation. Lemurs are as small-brained as typical mammals. Apes have bigger brains than monkeys, and humans have the biggest brains of all. The social brain hypothesis does not explain these variations. It sets up this problem: if social intelligence is so important, why do some group-living species have smaller brains than others?

Diet provides a major part of the answer. In 1995 Leslie Aiello and Peter Wheeler proposed that the reason some animals have evolved big brains is that they have small guts, and small guts are made possible by a high-quality diet. Aiello and Wheeler's head-spinning idea came from the realization that brains are exceptionally greedy for glucose—in other words, for energy. For an inactive person, every fifth meal is eaten solely to power the brain. Literally, our brains use around 20 percent of our basal metabolic rate—our energy budget when we are resting—even though they make up only about 2.5 percent of our body weight. Because human brains are so large, this proportion of energy expenditure is higher than it is in other animals: primates on average use about 13 percent of their basal metabolic rate on their brains, and most other mammals use less again, around 8 percent to 10 percent. As expected from the importance of maintaining energy flow to our many brain cells (neurons), genes that are responsible for energy metabolism show increased expression in the brains of humans compared to the brains of nonhuman primates. The high rate of energy flow is vital because our neurons need to keep firing whether we are awake or asleep. Even a brief interruption in the flow of oxygen or glucose causes neuron activity to stop, leading rapidly to death. The constant energy demand of brain cells continues even when times are tough, such as when food is scarce or an infection is raging. The first requirement for evolving a big brain is the ability to fuel it, and to do so reliably.

Given that large brains need large amounts of energy, Aiello and Wheeler asked themselves what special features of our species enable us to apportion more glucose to our brains than other animals do. One possibility is that humans might have a uniquely high rate of energy use. After all, human food is exceptionally calorie-dense and we routinely take in more energy per day than a typical primate of our body weight, so maybe extra energy running through our bodies gives us the calories we need to feed our hungry brains. But basal metabolic rates are well known in primates and other animals, and they are unremarkable in humans. A resting person supplies energy to their body at almost exactly the rate predicted for any primate of our body weight. Since nothing about basal metabolic rates is special to humans, Aiello and Wheeler were able to rule out the idea that our big brains are powered by inordinate amounts of energy passing through the body.

The elimination of the overall high-energy-use theory was a critical breakthrough because it left only one solution. Among species that have the same relative basal metabolic rate, such as humans and other primates, extra energy going to the brain must be offset by a reduced amount of energy going elsewhere. The question is what part of the body is shortchanged. Among primates, the size of most organs is closely predicted by body weight because of inescapable physiological rules. A species whose body weighs twice that of another needs a heart that weighs almost exactly twice as much. Hearts have to be a certain size to pump enough blood around a body of a certain size. No trade-off is possible there. Similar principles apply to kidneys, adrenals, and most other organs. But Aiello and Wheeler found a provocative exception to this tendency. They discovered that across the primates there is substantial variation in the relative weight of the intestinal system. Some species have big guts and some have small. The variation in gut size is linked to the quality of the diet.

Anyone who has handled tripe or cleaned a deer knows that mammals have a lot of gut tissue. Mammalian intestines have a high metabolic rate, and in large, mostly vegetarian species like great apes, intestines tend to be busy all day, starting with the postdawn meal and continuing ceaselessly until hours after the animal goes to sleep. All this time the guts are engaged in several energy-intensive functions, such as churning, making stomach acid, synthesizing digestive enzymes, or actively transporting digested molecules across the gut wall and into the blood. Active guts consume calories at a consistently high rate, so their total energy expenditure depends on their weight and on how much work they are doing. Carnivores, such as dogs and wolves, have smaller intestines than plant eaters, such as horses, cows, or antelope. In species that are adapted to eating more easily digested foods, such as sugar-rich fruits compared to fibrous leaves, guts are also relatively small: fruit-eating chimpanzees or spider monkeys have smaller guts than the leaf-eating gorillas or howler monkeys. Those reduced guts use less total energy than larger guts and therefore give a species with a high-quality diet some spare calories to allocate elsewhere in the body.

The discovery that gut size varies substantially gave Aiello and Wheeler the opening they were looking for. Relative to their body weight, primates with smaller guts proved to have larger brains—just the kind of trade-off that had been expected. Aiello and Wheeler estimated the number of calories a species is able to save by having a small gut, and showed that the number nicely matched the extra cost of the species' larger brains. The anthropologists concluded that primates that spend less energy fueling their intestines can afford to power more brain tissue. Big brains are made possible by a reduction in expensive tissue. The idea became known as the expensive tissue hypothesis.

Some species other than primates show a similar pattern, capitalizing on small guts to evolve particularly large brains. An elephant-nosed mormyrid fish from South America has a relatively tiny gut and is able to use an astonishing 60 percent of its energy budget to power its exceptionally large brain. Other animals follow the principle of an energy trade-off but gain muscle instead of brains. Birds that have small amounts of intestinal tissue tend to use their spare energy to grow bigger wing muscles, presumably because for a bird, better flight can be even more important than a bigger brain. Different kinds of trade-offs have also been proposed. Species with relatively low muscle mass have been found to have relatively large brains. The general lesson is that bigger brains must be paid for somehow. How animals with small guts make use of their energy savings depends on what matters to them. In primates the tendency to use energy saved by smaller guts for added brain tissue is particularly strong, presumably because most primates live in groups, where extra social intelligence has big payoffs.

The expensive tissue hypothesis predicted that major rises in human brain size would be associated with increases in diet quality. Aiello and Wheeler identified two such rises. The first brain-size expansion was around two million years ago from australopithecines to *Homo erectus*. In line with the

Man-the-Hunter scenario, the scientists credited this rise in brain size to the increased eating of meat. Second was a little more than half a million years ago, when *Homo erectus* became *Homo heidelbergensis*. They attributed this rise to the only other obvious candidate for an improvement in dietary quality: cooking.

I believe that Aiello and Wheeler were right in their principles. But they were wrong in their specifics because they assumed there was only a single increase in brain size from australopithecines to *Homo erectus*. In actuality, that phase of our evolution occurred in two steps: first, the appearance of the habilines, and second, the appearance of *Homo erectus*. Meat eating and cooking account respectively for these two transitions, and therefore for their accompanying increases in brain size.

The expensive tissue hypothesis provides an explanation not only for the substantial increases in brain size that occurred around the time of human origins, but also for the many other rises in brain size before and after two million years ago. Consider first our last common ancestor with chimpanzees, which lived around five million to seven million years ago. We can reconstruct this pre-australopithecine ape as living in rain forest and resembling a chimpanzee. Closely related to gorillas as well as chimpanzees, these ancestors likely had brains comparable in volume to those found in great apes living today, and therefore had larger brains than are found in living monkeys. The apes' big brains compared to those of monkeys are nicely explained by the expensive tissue hypothesis, because great apes have high-quality diets for their body weights. They eat relatively less fiber and fewer toxins than monkeys.

Chimpanzees have a cranial capacity of around 350 to 400 cubic centimeters (21.6 to 24.4 cubic inches). Australopithecines, with the same body weight as chimpanzees or even slightly less, had substantially larger cranial capacities, about 450 cubic centimeters (27.5 cubic inches). Following Aiello and Wheeler's hypothesis, australopithecine diets should therefore have been higher in quality than the diets of living chimpanzees. This seems likely. During seasons of plenty, australopithecines would have eaten much the same diet as chimpanzees or baboons do when living in the kinds of woodland that australopithecines occupied—fruits, occasional honey, soft seeds, and other choice plant items. It was when fruits were scarce that australopithecines must have eaten better than their chimpanzee-like ancestors. Present-day chimpanzees that are short of fruit turn to items specific to their rain-forest homes, eating foliage such as the stems of giant herbs and the soft young leaves of forest trees. In their drier woodlands australopithecines would have found few such items. The most likely alternatives were starch-filled roots and other underground or underwater storage tissues of herbaceous plants. These would have been ideal.

Carbohydrates are stored abundantly in corms, rhizomes, or tubers of many savanna plants and are highly concentrated sources of energy-rich starch in the dry season. These food reserves are so well hidden that few animals can find them, but chimpanzees do dig for tubers occasionally, sometimes with sticks, and australopithecines would have been at least as skillful and well-adapted: their chewing teeth are famously massive and somewhat piglike, suited to crushing roots and corms. An important location for australopithecine food sources likely would have been the edges of rivers and lakes, where sedges, water lilies, and cattails grow well and provide a natural supermarket of starchy foods for hunter-gatherers today.

The underground energy-storage organs of plants have a quality anticipated by the expensive tissue hypothesis: they have less indigestible fiber from plant cell walls than foliage, making them easier to digest and therefore a food of higher value. A dietary change from foliage to higher quality roots is thus a plausible explanation for the first increase in brain size, from forest apes to australopithecines five million to seven million years ago.

During the second sharp increase, brain volume rose by about one-third, from the roughly 450 cubic centimeters (27 cubic inches) of australopithecines to 612 cubic centimeters (37 cubic inches) in habilines (based on measurements of five skulls). The body weights of australopithecines and habilines were about the same, so this was a substantial gain in relative brain size. Given the archaeological evidence, the big dietary change at this time was more meat eating, so meat should have made this brain growth possible. To account for such a large increase in brain size, it seems likely that habilines processed their meat. Apes and humans are disadvantaged: their teeth cannot cut meat easily, their mouths are relatively small, and as William Beaumont noticed in the case of Alexis St. Martin, their stomachs do not process hunks of raw meat efficiently.

Chimpanzees also show that eating unprocessed meat is difficult with ape jaws. They chew their animal prey intensely, but small bits of undigested meat sometimes appear in their feces. Perhaps because of this hard work and inefficiency, chimpanzees sometimes decline the opportunity to eat meat despite their usual enormous enthusiasm for it. After chewing meat for an hour or two, a chimpanzee can abandon a carcass and relax or eat fruit instead. Chimpanzees of the Kanyawara community in Kibale National Park, Uganda, occasionally forgo meat-eating opportunities without chewing muscle at all. I once saw Johnny, an avid chimpanzee hunter of red colobus monkeys, do this even though he appeared hungry for animal protein. He first killed an infant red colobus monkey, brought it to the ground, ate its intestines, then left the carcass lying unseen by other chimpanzees. He immediately returned to the trees, rapidly killed another infant, and repeated his prior actions: he again brought his prey to the ground, ate the intestines, and left the rest to rot. His preference for the softer parts was typical. When chimpanzees kill a prey animal, they normally eat such parts as the guts, liver, or brain first. They can swallow those quickly. But when eating muscle, chimpanzees are forced to chew it slowly, taking as much as an hour to chew one-third of a kilogram (three-quarters of a pound). They can get as many calories per hour by chewing fruits as they can by chewing meat. The habilines would have faced the same challenge. If they had relied on unprocessed meat for as much as half their calories and had eaten their meat as slowly as chimpanzees, with certain cuts of meat they would have had to spend several hours a day chewing it. The digestive costs likewise would have been high, since the gut would have been busy digesting for many hours.

A system for hastening chewing and digestion by processing the meat would have greatly reduced the problem. Chimpanzees have a primitive form of processing meat. By adding tree leaves to their meat meals, they make chewing easier. The chosen leaves have no special nutritional properties, judging from the fact that the meat eaters pick leaves from whatever species of tree is nearest when they settle down to eat their prey. The only obvious rule governing their choice is that the leaf must be tough: they take only mature tree leaves, not young tree leaves or the soft leaves of an herb. Sometimes they even use long-dead leaves from the forest floor, mere brown skeletons devoid of nutrients. An informal experiment in which friends and I chewed raw goat meat suggested that the added leaves give traction. When we chewed thigh muscle together with a mature avocado leaf, the bolus of chewed meat was reduced faster than when we chewed with no added leaf. Australopithecines probably used similar practices when they caught gazelle fawns or other small mammals.

Habilines had access to more advanced techniques. Their bones are found close to stone hammers, fist-size spheres whose shapes provide vivid testimony of their repeated use. Habilines probably used the hammers partly to smash prey bones to extract the marrow. They also doubtless used the hammers to crack open nuts, as West African chimpanzees do, as well as to make other tools. In addition to these practices, stone hammers or wooden clubs could equally have been used for tenderizing meat. After habilines cut hunks of meat off the carcasses of game animals, they may have sliced them into steaks, laid them on flat stones, and pounded them with logs or rocks. Even relatively crude hammering would have reduced the costs of digestion by tenderizing the meat and breaking connective tissue. Because raw unprocessed meat is difficult to chew and digest, I suspect this was one of the most important cultural innovations in human origins, enabling habilines to increase the nutritional benefit of meat and the speed with which they could eat and digest it. Tenderizing meat would have reduced the costs of digestion by cutting the time that meat was in the stomach, and thus allowed habilines to divert energy toward their brains.

Dietary shifts toward roots, meat eating, and meat processing thus can explain the growth in brains from a chimpanzee-like ancestor at six million years to the habilines around two million years ago. From then on, the increases in brain size were more continuous. The habiline cranial capacity of 612 cubic centimeters (37 cubic inches) rose by over 40 percent to reach an average of 870 cubic centimeters (53 cubic inches) in the earliest measured *Homo erectus*. The significance of this rise is complicated by a parallel growth in body weight, from the lowly 32 to 37 kilograms (70 to 81 pounds) of habilines to a substantial 56 to 66 kilograms (123 to 145 pounds) in *Homo erectus*. Unfortunately, body weights are hard to estimate accurately from bones, and the number of specimens is small, so how much larger relative to body weight the brains of the first *Homo erectus* were than those of habilines, or whether they were relatively larger at all, is uncertain. However, *Homo erectus* brains continued to increase in size after 1.8 million years ago, averaging almost 950 cubic centimeters (58 cubic inches) by 1 million years ago. Given the evidence and arguments I have offered that *Homo erectus* originated as cooks, the expensive tissue hypothesis suggests their eating cooked food caused their brains to grow. Once cooking began, gut size could fall and the gut would be less active, both trends reducing the cost of the digestive system.

The fourth notable increase in cranial capacity occurred with the emergence of *Homo heidelbergensis* after eight hundred thousand years ago. The increase was again substantial, leading to a brain occupying around 1,200 cubic centimeters (73 cubic inches). This was the impressive rise that Aiello and Wheeler attributed to the invention of cooking—mistakenly, I believe. It remains a mystery, inviting speculation.

More efficient hunting is a possibility. Hartmut Thieme's evidence of group hunting four hundred thousand years ago in Schöningen suggests a marked improvement in hunting skills over earlier eras. This raises the possibility that meat intake, and perhaps therefore the use of animal fat, rose significantly before this time and played a role in the evolution of *Homo erectus* into *Homo heidelbergensis.*

Alternatively, cooking surely continued to affect brain evolution long after it was invented, because cooking methods improved. Laying a food item on the fire presumably was the main early method. Such techniques have been used by generations of campers and have been recorded by hunter-gatherers in recent times for foods that are easy to cook. The Aranda foragers of central Australia gather pea-size corms of sedges by digging them from flat ground near rivers. One method of cooking consists merely of laying them on hot ashes for a short time, then rubbing them between the hands to remove the light shell before eating them. !Kung San hunter-gatherers of Africa's Kalahari Desert cook tsin beans, one of their more important foods, by simply burying them in hot ashes. Putting an animal on a fire to roast can work fairly well, especially if the hairs have been singed off first. Marrow can be cooked with similar efficiency by roasting a complete bone in fire, then using stones to crack it. The marrow flows out like warm butter.

More complex ways to roast presumably would have accumulated slowly, often specific to particular foods. Take mongongo nuts eaten by !Kung hunter-gatherers. Mongongo nuts are a highly nutritious staple, often providing the !Kung with their major source of calories for weeks on end. To cook them, a woman mixes the coals from a dying fire with hot, dry sand. She then buries scores of nuts in the hot pile without allowing the nuts to touch any of the live coals. After a few minutes she kneads the pile to ensure that the nuts are evenly heated, adding more coals as needed. When the nuts are done, she hammers each one to split it, then eats the seeds inside or keeps them for further cooking. We do not know when such a sophisticated method appeared, but it seems likely to have contributed to raising the energetic quality of food, reducing the time the digestive system was active, and so lowering the total costs of digestion and allowing more energy for the brain.

Such improvements in cooking efficiency could explain why there was a steady upward trend in brain size during the lifetimes of the early human species. Brains were notably bigger in late *Homo erectus* than in early *Homo erectus*, and in late *Homo heidelbergensis* than in early *Homo heidelbergensis*. Major dietary breakthroughs such as meat eating and the invention of cooking cannot account for these smaller changes. The steady rise in brain size between the major jumps is most easily explained by a series of improvements in cooking techniques. Perhaps some particularly important advances enabled the prominent rise in brain size with *Homo heidelbergensis*.

The same possibility applies to the evolution of our own species, *Homo sapiens*, around two hundred thousand years ago. The gain in brain size was relatively minor, from 1,200 cubic centimeters (73 cubic inches) in *Homo heidelbergensis* to around 1,400 cubic centimeters (85 cubic inches) in *Homo sapiens*. Various modern behaviors are seen for the first time around this transition, such as the use of red ocher (presumably as a form of personal decoration), making tools out of bone, and long-distance trade. Increasing behavioral sophistication could also have happened in cooking techniques.

An early form of earth oven is the kind of innovation that could have been influential because it would have marked an important advance in cooking efficiency. Hunter-gatherers worldwide used earth ovens that employed hot rocks. The ovens do not appear to have been used by the people who expanded out of Africa more than sixty thousand years ago and colonized the rest of the world, since they are not recorded in Australia until thirty thousand years ago. However, it is possible that a simpler design, now vanished and forgotten, may have been used in earlier times.

In recent earth ovens the hot rocks provide an even, long-lasting heat. A typical procedure recorded in 1927 among the Aranda of central Australia involved digging a hole, filling it with a pile of dry wood, and topping that with large stones that did not crack when heated—often river cobblestones that had to be carried from a distance. When the stones were red-hot and fell through the fire, they were pulled out with sticks and the ashes were removed. The hot stones were then returned and covered with a layer of green leaves. Cooks liked to wrap meat in leaves to retain its juices before placing it on this layer, sometimes on top of a plant food such as roots. More green leaves and perhaps a basket mat would be laid on top, water was poured on, and some people added herbs for taste. Finally, the hole was filled with a layer of soil to retain the steam. After an hour or more—sometimes it was left overnight—the meat and vegetables would be ready and superb. The meat was laid on leafy branches, carved with a stone knife, and served. The even heat and moist environment made earth ovens efficient for gelatinizing starch and other carbohydrates, and they offered effective control over the tenderness of meat. This sophisticated cooking technique doubtless increased the digestibility of the meat and plant foods.

Likewise, the use of containers must have made cooking more efficient and might have contributed to reducing digestive costs and thus allowing increases in brain size. Pottery is a very recent invention, around ten thousand years ago, but natural objects could have been used as cooking containers long before that. Certain animals come with their own dishes. Shellfish, such as mussels, have been cooked whole in many parts of the world by being thrown into a fire until the valves open. The Yahgan of Tierra del Fuego used mussel shells to catch the drips from a roasting seal or to hold whale oil, which they ate by dipping pieces of edible fungus into it.

It is a small step from such techniques to cooking in a container. Heating in natural containers by early *Homo sapiens* is indicated around 120,000 years ago by evidence that people made a glue from ancient birch tar, which they used to haft stone points on to spears. The glue had to be heated to achieve the desired stickiness, so people must have been cooking with containers by then. Some containers would have needed little imagination. Turtles are a natural convenience food because they can be easily kept alive for days, and whether alive or cooked they are easily carried. If they are turned upside down they even provide their own cooking pot. After their flesh has been eaten, their bodies remain useful. Andaman Islanders from the Bay of Bengal cooked turtle blood in an upside-down shell until it was thick, then ate it at once. Like many Asian peoples they also used bamboo as a container, sometimes for cooking. The Andaman Islanders would clean a length of bamboo and heat it over a fire so all its juices were absorbed. They then packed it with half-cooked pieces of wild pork or other meat and heated it so slowly that the meat swelled without cracking the bamboo. When the bamboo stopped steaming, they removed it from the fire and stuffed the opening with leaves to seal it. The cooked meat could be left for several days. Sadly, many ingenious cooking techniques practiced by early people with plant materials are forever lost to us because they leave no traces.

Development of other methods would have improved the efficiency of cooking and the quality of food. Various special ways of roasting have unknown antiquity. In their cold climate near the Antarctic, the Yahgan developed a two-stone griddle by heating two flat stones in a fire. The stones were then withdrawn, and the larger stone served as a griddle for a steak or layer of blubber, while the smaller was laid on top. This worked so well that the fat was browned and shriveled in a few minutes, a favorite for the hunters. The Yahgan were also fond of sausages. To make a sea-lion blood sausage, they kept the blood that collects in the abdominal cavity of a freshly killed sea lion. They took a soft, still moist piece of gut, turned it inside out, cleaned it, tied it shut at one end with sinews, filled it with air by blowing, tied the other end shut, and left it to dry. When the empty sausage was sufficiently firm, they used a large shell to fill it with blood, tied it shut again, and for safety's sake jabbed a short, thin stick into each end to prevent the ties from unraveling. They then put the sausage into hot ashes, poking it occasionally to keep it from bursting. The same idea worked equally well with other parts of the gut. They sometimes filled stomachs with blubber or chopped tissues like heart, lungs, or liver. These haggises of the past left no traces, but they remind us that even in the bush, long before such recent inventions as grinding and stone boiling (which started within the past twenty-five thousand to forty thousand years), cooking can involve much more than simple heating.

Although the breakthrough of using fire at all would have been the biggest culinary leap, the subsequent discovery of better ways to prepare the food would have led to continual increases in digestive efficiency, leaving more energy for brain growth. The improvements would have been especially important for brain growth after birth, since easily digested weaning foods would have been critical contributors to a child's energy supply. Advances in food preparation may thus have contributed to the extraordinary continuing rise in brain size through two million years of human evolution—a trajectory of increasing brain size that has been faster and longer-lasting than known for any other species. When Charles Darwin called cooking "probably the greatest [discovery], excepting language, ever made by man," he was thinking merely of our improved food supply. But the idea that brain enlargement was made possible by improvements in diet suggests a wider significance. Cooking was a great discovery not merely because it gave us better food, or even because it made us physically human. It did something even more important: it helped make our brains uniquely large, providing a dull human body with a brilliant human mind.

CHAPTER 6

How Cooking Frees Men

"Voracious animals . . . both feed continually and as incessantly eliminate, leading a life truly inimical to philosophy and music, as Plato has said, whereas nobler and more perfect animals neither eat nor eliminate continually."

—GALEN, *Galen on the Usefulness of the Parts of the Body*

Diet has long been considered a key to understanding social behavior across species. The food quest is fundamental to evolutionary success, and social strategies affect how well individuals eat. Group size in chimpanzees rapidly adjusts to monthly changes in the density and distribution of fruiting trees. Chimpanzee society differs markedly from gorilla society, thanks to the gorillas' reliance on herbs. Humans are no exception to such relationships. The Man-the-Hunter hypothesis has inspired such potent explanations of bonding between males and females that it has seemed to some researchers that no other explanation is necessary. In 1968 physical anthropologists Sherwood

Washburn and Chet Lancaster wrote, "Our intellect, interests, emotions and basic social life, all are evolutionary products of the hunting adaptation." Such ideas have been highly influential, but they have rarely looked beyond meat. The adoption of cooking must have radically changed the way our ancestors ate, in ways that would have changed our social behavior too.

Take softness. Foods soften when they are cooked, and as a result, cooked food can be eaten more quickly than raw food. Reliance on cooked food has therefore allowed our species to thoroughly restructure the working day. Instead of chewing for half of their time, as great apes tend to do, women in subsistence societies tend to spend the active part of their days collecting and preparing food. Men, liberated from the simple biological demands of a long day's commitment to chewing raw food, engage in productive or unproductive labor as they wish. In fact, I believe that cooking has made possible one of the most distinctive features of human society: the modern form of the sexual division of labor.

The sexual division of labor refers to women and men making different and complementary contributions to the household economy. Though the specific activities of each sex vary by culture, the gendered division of labor is a human universal. It is therefore assumed to have appeared well before modern humans started spreading across the globe sixty thousand to seventy thousand years ago. So discussion of the evolution of the sexual division of labor centers on hunter-gatherers. The 750-strong Hadza tribe are one such group. They live in northern Tanzania, scattered among a series of small camps in dry bush country around a shallow lake.

The Hadza are modern-day people. Neighboring farmers and pastoralists trade with them and marry some of their daughters. Government officials, tourists, and researchers visit them. The Hadza use metal knives and money, wear cotton clothes, hunt with dogs, and occasionally trade for agricultural foods. Much has changed since the time, perhaps two thousand years ago, when they last lived in an exclusive world of hunter-gatherers. Nevertheless, they are one of the few remaining peoples who obtain the majority of their food by foraging in an African woodland of a type that was once occupied by ancient humans.

Dawn sees people emerging from their sleeping huts to eat scraps of food from the previous night's meal. As consensus quietly develops about the day's activity; most of the women in camp—six or more, perhaps—take up their digging sticks and go toward a familiar *ekwa* patch a couple of kilometers (more than a mile) away. Some take their babies in slings, and one or more carries a smoldering log with which to start a fire if needed. Older children walk alongside. Meanwhile, in ones and twos, various men and their dogs also walk off with bows and arrows in hand. Some men are going hunting, others to visit neighbors. A scattering of people remain in camp—a couple of old women, perhaps, looking after toddlers whose mothers have gone for food, and a young man resting after a long hunt the previous day.

The women walk slowly, in pace with the younger children. They stop occasionally to pick small fruits that they eat on the spot. After less than an hour they break into smaller parties as each forager finds her own choice site in calling distance of her companions. The digging is hard and uncomfortable but it does not take long. A couple of hours later the women's karosses—cloaks made of animal skins—are covered in piles of thick, brown, foot-long roots. These *ekwa* tubers are a year-round staple for the Hadza, always easily found. As the karosses fill, someone starts a fire, and shortly afterward the foragers gather for a well-deserved snack. They bake their *ekwa* by leaning the tubers against the coals. In barely twenty minutes, the smaller ones are ready. After the simple meal, some women chat while others dig up a few more *ekwa* to make sure they have enough for the rest of the day. Most have found other foods as well—a few bulbs, perhaps. They tie up their karosses and start homeward. Each woman totes at least 15 kilograms (33 pounds). They are back in camp by early afternoon, tired from the hard work.

Anthropologists sometimes debate whether hunting and gathering is a relaxed way of life. Lorna Marshall worked alongside Nyae Nyae !Kung women gathering in the Kalahari in the 1950s. "They did not have pleasurable satisfaction," she said, "in remembering their hot, monotonous, arduous days of digging and picking and trudging home with their heavy loads." But times and cultures vary. Anthropologist Phyllis Kaberry, who worked with aborigines in the Kimberley region of northwestern Australia, said the women enjoyed one another's company and their foraging routine.

Back in the Hadza camp, each woman empties her kaross in her own hut. By early evening she has a fire, and a pile of *ekwa* lies baked and ready. She hopes the men will bring some meat to complete the meal. During the evening hours several men return. Some have honey, a few have nothing, and one arrives with the carcass of a warthog. After he singes the animal's hair off in a fire, men and women gather to divide it. Following the typical practice of hunter-gatherers, many men in the camp get a share, but the successful hunter makes sure his friends, family, and relatives get the most. Soon each household fire is cooking meat. The delicious smells enrich the night air. The meat and the roasted *ekwa* are quickly consumed. As the camp settles into sleep, enough *ekwa* remains for breakfast the following day.

The Hadza illustrate two major features of the sexual division of labor among hunter-gatherers that differentiate humans sharply from nonhuman primates. Women and men spend their days seeking different kinds of foods, and the foods they obtain are eaten by both sexes. Why our species forages in such an unusual way (compared to primates and all other animals, whose adults do not share food with one another) has never been fully resolved. There are many variations in the particular foods obtained. Tierra del Fuego's bitter climate provided few plant foods, so while men hunted sea mammals, women would dive for shellfish in the frigid shallows. In the tropical islands of northern Australia, there was so much plant food that women brought enough to feed all the family and still found time to hunt occasional small animals. Men there did little hunting, mostly playing politics instead.

Although the specific food types varied from place to place, women always tended to provide the staples, whether roots, seeds, or shellfish. These foods normally needed processing, which could involve a lot of time and laborious work. Many Australian tribes prepared a kind of bread called damper from small seeds, such as from grasses. Women gathered the plants and heaped them so their seeds would drop and collect in a pile. They threshed the seeds by trampling, pounding, or rubbing them in their hands, winnowed them in long bark dishes, and ground them into a paste. The result was occasionally eaten raw but was more often cooked on hot ashes. The whole process could take more than a day. Women worked hard at such tasks because their children and husbands relied on the staples women prepared.

Men, by contrast, tended to search for foods that were especially appreciated but could not be found easily or predictably. They hoped for such prizes as meat and honey, which tended to come in large amounts and tasted delicious. Their arrival in camp made the difference between happiness and sadness. Phyllis Kaberry's description of an aborigine camp in western Australia is typical: "The Aborigines continually craved for meat, and any man was apt to declare, 'me hungry alonga bingy,' though he had had a good meal of yams and damper a few minutes before. The camp on such occasions became glum, lethargic, and unenthusiastic about dancing." Hunting large game was a predominantly masculine activity in 99.3 percent of recent societies.

Hints of comparable sex differences in food procurement have been detected in primates. Female lemurs tend to eat more of the preferred foods than males. In various monkeys such as macaques, guenons, and mangabeys, females eat more insects and males eat more fruit. Among chimpanzees, females eat more termites and ants, and males eat more meat. But such differences are minor because in every nonhuman primate the overwhelming majority of the foods collected and eaten by females and males are the same types.

Even more distinctive of humans is that each sex eats not only from the food items they have collected themselves, but also from their partner's finds. Not even a hint of this complementarity is found among nonhuman primates. Plenty of primates, such as gibbons and gorillas, have family groups. Females and males in those species spend all day together, are nice to each other, and bring up their offspring together, but, unlike people, the adults never give each other food. Human couples, by contrast, are expected to do so.

In foraging societies a woman always shares her food with her husband and children, and she gives little to anyone other than close kin. Men likewise share with their wives, whether they have received meat from other men or have brought it to camp themselves and shared part of it with other men. The exchanges between wife and husband permeate families in every society. The contributions might involve women digging roots and men hunting meat in one culture, or women shopping and men earning a salary in another. No matter the specific items each partner contributes, human families are unique compared to the social arrangements of other species because each household is a little economy.

Attempts to understand how the sexual division of labor arose in our evolutionary history have been strongly affected by whether women or men are thought to have provided more of the food. It used to be thought that women typically produced most of the calories, as occurs among the Hadza. Worldwide across foraging groups, however, men probably supplied the bulk of the food calories more often than women did. This is particularly true in the high, colder latitudes where there are few edible plants, and hunting is the main way to get food. In an analysis of nine well-studied groups, the proportion of calories that came from foods collected by women ranged from a maximum of 57 percent, in the desert-living G/wi Bushmen of Namibia, down to a low of 16 percent in the Aché Indians of Paraguay. Women provided one-third of the calories in these societies, and men two-thirds. But such averages do not give an accurate sense of the value of items each sex contributes. At different times of year, the relative importance of foods obtained by women and men can change, and overall each sex's foods can be just as critical as the other's in maintaining health and survival. Furthermore, each sex makes vital contributions to the overall household economy regardless of any difference in the proportion of food calories contributed.

The division of labor by sex affects both household subsistence and society as a whole. Sociologist Emile Durkheim thought that its most important result was to promote moral standards, by creating a bond within the family. Specialization of labor also increases productivity by allowing women and men to become more skilled at their particular tasks, which promotes efficient use of time and resources. It is even thought to be associated with the evolution of some emotional and intellectual skills, because our reliance on sharing requires a cooperative temperament and exceptional intelligence. For such reasons anthropologists Jane and Chet Lancaster described the sexual division of labor as the "fundamental platform of behavior for the genus *Homo*," and the "true watershed for differentiating ape from human lifeways." Whether they were right in thinking the division began with the genus *Homo* is debated. Though I agree with the Lancasters, many think the division of labor by sex started much later. But there is no doubt of its importance in making us who we are.

The classic explanation in physical anthropology for this social structure is essentially what Jean Anthelme Brillat-Savarin proposed: when meat became an important part of the human diet, it was harder for females than males to obtain. Males with a surplus would have offered some to females, who would have appreciated the gift and returned the favor by gathering plant foods to share with males. The result was an incipient household. Physical anthropologist Sherwood Washburn put it this way:

> When males hunt and females gather, the results are shared and given to the young, and the habitual sharing between a male, a female, and their offspring becomes the basis for the human family. According to this view, the human family is the result of the reciprocity of hunting, the addition of a male to the mother-plus-young social group of the monkeys and apes.

Washburn's statement captures a core feature of conventional wisdom, which is that the way to explain the evolution of the sexual division of labor is to imagine that, together, meat eating and plant eating allowed a household. An unstated assumption was that the food was raw. But if food was raw, the sexual division of labor is unworkable. Nowadays a man who has spent most of the day hunting can satisfy his hunger easily when he returns to camp, because his evening meal is cooked. But if the food waiting for him in camp had all been raw, he would have had a major problem.

The difficulty lies in the large amount of time it takes to eat raw food. Great apes allow us to estimate it. Simply because they are big—30 kilograms (66 pounds) and more—they need a lot of food and a lot of time to chew. Chimpanzees in Gombe National Park, Tanzania, spend more than six hours a day chewing. Six hours may seem high considering that most of their food is ripe fruit. Bananas or grapefruit would slip down their throats easily, and for this reason, chimpanzees readily raid the plantations of people living near their territories. But wild fruits are not nearly as rewarding as those domesticated fruits. The edible pulp of a forest fruit is often physically hard, and it may be protected by a skin, coat, or hairs that have to be removed. Most fruits have to be chewed for a long time before the pulp can be fully detached from the pieces of skin or seeds, and before the solid pieces are mashed enough to give up their valuable nutrients. Leaves, the next most important food for chimpanzees, are also tough and likewise take a long time to chew into pieces small enough for efficient digestion. The other great apes (bonobos, gorillas, and orangutans) commit similarly long hours to chewing their food. Because the amount of time spent chewing is related to body size among primates, we can estimate how long humans would be obliged to spend chewing if we lived on the same kind of raw food that great apes do. Conservatively, it would be 42 percent of the day, or just over five hours of chewing in a twelve-hour day.

People spend much less than five hours per day chewing their foods. Brillat-Savarin claimed to have seen the vicar of Bregnier eat the following within forty-five minutes: a bowl of soup, two dishes of boiled beef, a leg of mutton, a handsome capon, a generous salad, a ninety-degree wedge from a good-sized white cheese, a bottle of wine, and a carafe of water. If Brillat-Savarin was not exaggerating, the amount of food eaten by the vicar in less than an hour would have provided enough calories for a day or more. It is hard to imagine a wild chimpanzee achieving such a feat.

A few careful studies using direct observation confirm how relatively quickly humans eat their food. In the United States, children from nine to twelve years of age spend a mere 10 percent of their time eating, or just over an hour per twelve-hour day. This is close to the daily chewing time for children recorded by anthropologists in twelve subsistence societies around the world, from the Ye'kwana of Venezuela to the Kipsigi of Kenya and the Samoans of the South Pacific. Girls ages six to fifteen chewed for an average of 8 percent of the day, with a range of 4 percent to 13 percent. Results for boys were almost identical: they chewed for an average of 7 percent of the day, again ranging from 4 percent to 13 percent.

The children's data show little difference between the industrialized United States and subsistence societies. In the twelve measured cultures, adults chewed for even less time than the children. Women and men each spent an average of 5 percent of their time chewing. One might object that the people in the subsistence societies were observed only from dawn to dusk. Since people often have a big meal after dark, the total time eating per day might be more than indicated by the 5 percent figure, which translates to only thirty-six minutes in a twelve-hour day. But even if people chewed their evening meals for an hour after dark, which is an improbably long time, the total time spent eating would still be less than 12 percent of a fourteen-hour day, allowing two hours for the evening meal. However we look at the data, humans devote between a fifth and a tenth as much time to chewing as do the great apes.

This reduction in chewing time clearly results from cooked food being softer. Processed plant foods experience similar physical changes to those of meat. As the food canning industry knows all too well, it is hard to retain a crisp, fresh texture in heated vegetables or fruits. Plant cells are normally glued together by pectic polysaccharides. These chemicals degrade when heated, causing the cells to separate and permitting teeth to divide the tissue more easily. Hot cells also lose rigidity, a result of both their walls swelling and their membranes being disrupted by denaturation of proteins. The consequences are predictable. By measuring the amount of force needed to initiate a crack in food, researchers have shown that softness (or hardness) closely predicts the number of times someone chews before swallowing. The effect works for animals too. Wild monkeys spend almost twice as long chewing per day if their food is low-quality. Observers have recorded the amount of time spent chewing by wild primates that obtain human foods (such as garbage stolen from hotels). As the proportion of human foods rises in the diet, the primates spend less time chewing, down to less than 10 percent when all of the food comes from humans.

Six hours of chewing per day for a chimpanzee mother who consumes 1,800 calories per day means that she ingests food at a rate of around 300 calories per hour of chewing. Humans comparatively bolt their food. If adults eat 2,000 to 2,500 calories a day, as many people do, the fact that they chew for only about one hour per day means that the average intake rate will average 2,000 to 2,500 calories an hour or higher, or more than six times the rate for a chimpanzee. The rate is doubtless much more when people eat high-calorie foods, such as hamburgers, candy bars, and holiday feasts. Humans have clearly had a long history of much more intense calorie consumption than primates are used to. Thanks to cooking, we save ourselves around four hours of chewing time per day.

Before our ancestors cooked, then, they had much less free time. Their options for subsistence activities would therefore have been severely constrained. Males could not afford to spend all day hunting, because if they failed to get any prey, they would have had to fill their bellies on plant foods instead, which would take a long time just to chew. Consider chimpanzees, who hunt little and whose raw-food diet can be safely assumed to be similar to the diet of australopithecines. At Ngogo, Uganda, chimpanzees hunt intensely compared to other chimpanzee populations, yet males still average less than three minutes per day hunting. Human hunters have lots of time and walk for hours in the search for prey. A recent review of eight hunter-gatherer societies found that men hunted for between 1.8 and 8.2 hours daily. Hadza men were close to the average, spending more than 4 hours a day hunting—about eighty times as long as an Ngogo chimpanzee.

Almost all hunts by chimpanzees follow a chance encounter during such routine activities as patrolling their territorial boundaries, suggesting that chimpanzees are unwilling to risk spending time on a hopeful search. When chimpanzees hunt their favorite prey—red colobus monkeys—the colobus rarely move out of the tree where they are attacked. The monkeys appear to feel safer staying in one place, rather than jumping to adjacent trees where chimpanzees might ambush them. The monkeys' immobility allows chimpanzees to alternate between sitting under the prey and making repeated rushes at them. In theory, the chimpanzees could spend hours pursuing this prey. But at Ngogo the longest hunt observed was just over one hour, and the average length of hunts is only eighteen minutes. At Gombe I found that the average interval between plant-feeding bouts was twenty minutes, almost the same as the length of a hunt. The similarity between the average hunt duration and the average interval between plant-feeding bouts suggests that chimpanzees can afford a break of twenty minutes from eating fruits or leaves to hunt, but if they take much longer they risk losing valuable plant-feeding time.

The time budget for an ape eating raw food is also constrained by the rhythm of digestion, because apes have to pause between meals. Judging from data on humans, the bigger the meal, the longer it takes for the stomach to empty. It probably takes one to two hours for a chimpanzee's full stomach to empty enough to warrant feeding again. Therefore, a five-hour chewing requirement becomes an eight- or nine-hour commitment to feeding. Eat, rest, eat, rest, eat. An ancestor species that did not cook would presumably have experienced a similar rhythm.

These time constraints are inescapable for a large ape or habiline eating raw unprocessed food. Males who did not cook would not have been able to rely on hunting to feed themselves. Like chimpanzees, they could hunt in opportunistic spurts. But if they devoted many hours to hunting, the risk of failure to obtain prey could not be compensated rapidly enough. Eating their daily required calories in the form of their staple plant foods would have taken too long.

Washburn and other anthropologists have proposed that the human division of labor by sex was based on hunting. They suggest that on days when a male failed to find meat, honey, or other prizes, a female could provide food to him. As we now see, this would not have been sufficient, because a returning male who had not eaten during the day would not have had enough time left in the evening to chew his plant-food calories. The same time constraints apply whether our precooking ancestor obtained his staple plant diet by his own labor or received it from a female. A division of labor into hunting and gathering would not have afforded consumption of sufficient calories, as long as the food was consumed raw.

Suppose that a hunter living on raw food has a mate who is willing to feed him, that his mate could collect enough raw foods for him (while satisfying her own needs) and would bring them back to a central place, to be met by her grateful mate. Then suppose the male has had an unsuccessful day of hunting. Even modern hunter-gatherers armed with efficient weapons often fail. Among the Hadza there are stretches of a week or more several times per year when hunters bring no big-game meat to camp. The hungry hunter needs to consume, say, two thousand calories, but he cannot eat after dark. To do so would be too dangerous, scrabbling in the predator-filled night to feel for the nuts, leaves, or roots his gatherer friend brought him. If the hunter slept on the ground, he would be exposed to predators and large ungulates as he fumbled for his food. If he were in a tree, he would find it hard to have his raw foods with him because they do not come in tidy packages.

So to eat his fill he would have to do most of his eating before dusk, which falls between about 6 and 7 P.M. in equatorial regions. If he had eaten nothing while on the hunt, he would need to be back in camp before midday, and there he would find his mate's gathered foods (assuming she had been able to complete her food gathering so early in the day). He would then have to spend the rest of the day eating, resting, eating, resting, and eating. In short, the long hours of chewing necessitated by a raw diet would have sharply reduced hunting time. It is questionable whether the sexual division of labor would have been possible at all.

The use of fire solved the problem. It freed hunters from previous time constraints by reducing the time spent chewing. It also allowed eating after dark. The first of our ancestral line to cook their food would have gained several hours of daytime. Instead of being an opportunistic activity, hunting could have become a more dedicated pursuit with a higher potential for success. Nowadays men can hunt until nightfall and still eat a large meal in camp. After cooking began, therefore, hunting could contribute to the full development of the family household, reliant as it is on a predictable economic exchange between women and men.

CHAPTER 7

The Married Cook

"The labor of women in the house, certainly, enables men to produce more wealth than they otherwise could; and in this way women are economic factors in society. But so are horses . . . the horse is not economically independent, nor is the woman."

—CHARLOTTE PERKINS GILMAN,
Women and Economics: A Study of the Economic Relation Between Men and Women as a Factor in Social Evolution

An evening meal cooked by a woman serves her and her children's needs. It also helps her husband by giving him a predictable source of food, allowing him to spend his day doing whatever activity he chooses. But while the arrangement is comfortable for both sexes, it is particularly convenient for the male. Why should a female cook for him? A focus on the peculiar properties of cooked food provokes a new understanding of the nature of married life and the human community. It suggests that the reasons why the sexes pair off go beyond the traditional ideas of mating competition, or the interests that women and men have in the product of each other's labor. It leads to the uncomfortable idea that as a cultural norm, women cook for men because of patriarchy. Men use their communal power to consign women to domestic roles, even when women would prefer otherwise.

That women tend to cook for their husbands is clear. In 1973 anthropologists George Murdock and Catarina Provost compiled the pattern of sex differences in fifty productive activities in 185 cultures. Although men often like to cook meat, overall cooking was the most female-biased activity of any, a little more so than preparing plant food and fetching water. Women were predominantly or almost exclusively responsible for cooking in 97.8 percent of societies. There were four societies in which cooking was reportedly performed about equally by both sexes or predominantly by males. One of them, the Todas of South India, was an error: a 1906 report had been misleading. Murdock and Provost failed to catch a correction showing that Toda women did most of the cooking.

Even the apparent exceptions conformed to the general rule. The three standouts reveal an important distinction between two types of cooking: cooking for the family, done by women, and cooking for the community, done by men. The three were the Samoans, Marquesans, and Trukese, all in the South Pacific. Their cultural backgrounds are different and they are located hundreds of miles apart from one another, but they have one thing in common: their staple food is breadfruit. Breadfruit trees produce fruits the size of basketballs, yield large volumes of high-quality starch, and demand cooperative processing.

The procedure for cooking the fruit pulp is physically arduous, takes many hours, and is performed in a communal house by a group of men on days of their choice. The men build a large fire, peel the fruits, cut them into chunks, and steam them. On the Truk island group of Micronesia (now often called Chuuk), the ringing sound of sweat-soaked men pounding the fruit meat with coral pestles could be heard a hundred yards away. It was late in the day before men were done wrapping the cooled mash into leaf packages. They distributed the surplus to men who had not been cooking. At the end of the day, all the men had food packages, and sometimes they ate together in the men's house, where women were not allowed.

Men did not need women to feed them. They could spend weeks at a time in the men's houses with men of their lineage, receiving no assistance from women. But when men ate at home they gave the breadfruit mash to their wives, and their wives used it as the basis for the evening meal. Women rounded it out with pork or fish sauces and vegetables they cooked themselves. If there was no breadfruit, the women cooked other starchy foods such as taro roots. Men cooked the main staple when they chose to do so, but women were responsible for cooking everything else and for producing household meals.

Might there be a few societies, not sampled by Murdock and Provost, in which women are so liberated that the gendered pattern of cooking is reversed? Cultural anthropologist Maria Lepowsky studied the people of Vanatinai in the South Pacific expressly because, from the outside, this society seemed like a woman's dream community. In many ways, life was indeed very good for women. There was no ideology of male superiority. Both sexes could host feasts, lead canoeing expeditions, raise pigs, hunt, fish, participate in warfare, own and inherit land, decide about clearing land, make shell necklaces, and trade in such valued items as greenstone ax blades. Women and men were equally capable of attaining the prestige of being "big" (important) people. Domestic violence was rare and strongly censured. There was "tremendous overlap in the roles of men and women" and a great deal of personal control over how they chose to spend their time. Women had "the same kinds of personal autonomy and control of the means of production as men."

Yet despite the apparent escape from patriarchy, women on Vanatinai did all the domestic cooking. Cooking was regarded as a low-prestige activity. Other chores for which women were responsible included washing dishes, fetching water and firewood, sweeping, and cleaning up pig droppings. All were again regarded as low-status duties—in other words, the kind of work men did not want to do. One day as a group of women returned after walking three miles with heavy baskets of yams on their heads, they complained to

Lepowsky, "We come home after working in the garden all day, and we still have to fetch water, look for firewood, do the cooking and cleaning up and look after the children while all men do is sit on the verandah and chew betel nut!" But when they asked for help with these tasks, "The men," wrote Lepowsky, "usually retort that these are the work of women." Why should the men help, if they can get away with not helping?

The worldwide pattern is reflected in the English language. The word *lady* is derived from the Old English *hlaefdige*, meaning "bread kneader," whereas *lord* comes from *hlaefweard*, or "bread keeper." Of course, men are entirely capable of cooking. In industrial societies men can be professional cooks. Spouses in urban marriages often share the cooking, or husbands can do most of it. In hunter-gatherer societies men cook for themselves on long hunting expeditions or in bachelor groups. Men cook on feast days and ritual occasions, cooperating in public somewhat like the breadfruit cooks. But even the men who cook when no women are present or on ceremonial occasions still have their home foods prepared by women. The rule that domestic cooking is women's work is astonishingly consistent.

The classic reason suggested for this pattern is mutual convenience. Each sex gains from sharing their efforts, as many happily married couples can attest. But the explanation is superficial because it does not address the more fundamental problem of why our species has households at all, or the darker dynamic that sometimes has husbands exploiting their wives' labor. The men on Vanatinai could have shared the cooking easily, as the women would sometimes have liked them to do, but they chose not to. Charlotte Perkins Gilman noted that humans are the only species in which "the sex-relation is also an economic relation" and compared women's role to that of horses. Molly and Eugene Christian complained that cooking "has made of woman a slave." In theory, among hunter-gatherers both males and females could forage for themselves, like every other animal, and then cook his or her own meal at the end of the day. So what led to a sexual division of labor in which men routinely insist that it is women's lot to do the household cooking?

Nonhuman primates mostly pick and eat their food at once. But hunter-gatherers bring food to a camp for processing and cooking, and in the camp, labor can be offered and exchanged. This suggests that cooking might be responsible for converting individual foraging into a social economy. Archaeologist Catherine Perlès thinks so: "The culinary act is from the start a project. Cooking ends individual self-sufficiency." Relying on cooking creates foods that can be owned, given, or stolen. Before cooking, we ate more like chimpanzees, everyone for themselves. After the advent of cooking, we assembled around the fire and shared the labor.

Perlès's notion that, by necessity, cooking was a social activity is supported by Dutch sociologist and fire expert Joop Goudsblom, who suggests that cooking required social coordination, "if only to ensure that there would always be someone to look after the fire." Food historian Felipe

Fernandez-Armesto proposed that cooking created mealtimes and thereby organized people into a community. For culinary historian Michael Symons, cooking promoted cooperation through sharing, because the cook always distributes food. Cooking, he wrote, is "the starting-place of trades."

These ideas fit nicely with the ubiquitous social importance of cooked food. The contrast between communal and solitary eating is particularly pronounced among hunter-gatherers, for whom cooking is a highly social act, unlike eating raw food. When people are out of camp, their snacks tend to be raw foods such as ripe fruits or grubs, and these are normally collected individually and eaten without sharing. But when people cook food, they do so mostly in camp, and they share it within the family or, when feasting, with other families. Furthermore, much of the labor in preparing the meal is complementary. In a common pattern, a woman brings firewood and vegetables, prepares the vegetables, and does the cooking. A man brings meat, which either he or a woman might cook. Family members also tend to eat at roughly the same time (though the man may eat first) and often sit face to face around a fire.

But the suggestion that tending a fire, eating a meal, and sharing food require cooperation is obviously wrong. Alexander Selkirk, the real-life Robinson Crusoe, was very fit when he was rescued in 1709 after more than four years of cooking for himself in the Juan Fernandez Islands in the middle of the Pacific. Numerous solitary war survivors also have lived off the land and cooked for themselves, as Shoichi

Yokoi did in Guam for almost thirty years before he was found in 1972. Hunter-gatherer women sometimes collect food and fuel, tend a fire, and do the cooking without any support from their husbands, such as Tiwi women in northern Australia. Men in societies ranging from hunter-gatherers to the United States can go on hunting expeditions for days at a time and cook for themselves. Examples of individual self-sufficiency clearly undermine the idea that the sheer mechanics of cooking require that it be practiced cooperatively.

Why, then, is the "culinary project" so often social, if it does not need to be? Relying on cooked food creates opportunities for cooperation, but just as important, it exposes cooks to being exploited. Cooking takes time, so lone cooks cannot easily guard their wares from determined thieves such as hungry males without their own food. Pair-bonds solve the problem. Having a husband ensures that a woman's gathered foods will not be taken by others; having a wife ensures the man will have an evening meal. According to this idea, cooking created a simple marriage system; or perhaps it solidified a preexisting version of married life that could have been prompted by hunting or sexual competition. Either way, the result was a primitive protection racket in which husbands used their bonds with other men in the community to protect their wives from being robbed, and women returned the favor by preparing their husbands' meals. The many beneficial aspects of the household, such as provisioning by males, increases in labor efficiency, and creation of a social network for child-rearing, were additions consequent to solving the more basic problem: females needed male protection, specifically because of cooking. A male used his social power both to ensure that a female did not lose her food, and to guarantee his own meal by assigning the work of cooking to the female.

The logic for this theory begins with the banal observation that cooking is necessarily a conspicuous and lengthy process. In the bush, the sight or smell of smoke reveals a cook's location at a long distance, allowing hungry individuals who have no food to easily locate cooks in action. The effect among *Homo erectus* is easily imagined. Because females were smaller and physically weaker, they were vulnerable to bullying by domineering males who wanted food. Each female therefore obtained protection from other males' wheedling, scrounging, or bullying by forming a special friendship with her own particular male. Her bond with him protected her food from other males, and he also gave her meat. These bonds were so critical for the successful feeding of both sexes that they generated a particular kind of evolutionary psychology in our ancestors that shaped female-male relationships and continues to affect us today.

The idea that cooking has influenced social relationships in this way is supported by the intense aversion to competition shown by hunter-gatherers eating their meals. Lorna Marshall's description of the delicacy with which Nyae Nyae !Kung treat one another at mealtimes is typical of hunter-gatherers: "When a visitor comes to the fire of a family which is preparing food or eating, he should sit at a little distance, not to seem importunate, and wait to be asked to share. . . . We observed no unmannerly behavior and no cheating and no encroachment about food. . . . The polite way to receive food, or any gift, is to hold out both hands and have the food or other gift placed in them. To reach out with one hand suggests grabbing to the !Kung. I found it moving to see so much restraint about taking food among people who are all thin and often hungry, for whom food is a constant source of anxiety."

Such spontaneous etiquette is universal within functioning hunter-gatherer societies. Nothing like it is found in any other social species. Among nonhuman animals, valuable items that cannot be eaten at once predictably induce fights. Most of the fruits eaten by chimpanzees are the size of plums or smaller, too small to be worth fighting over, but a single ripe breadfruit weighs up to eight kilograms (eighteen pounds) and can take a group two hours to eat. An individual does not have time to swallow it before others see the prize and come to compete for it. Offspring take advantage of the situation by begging from their mothers, and adults fight to possess whole fruits or large pieces. Among chimpanzees, males win. Among bonobos, females win. In each case, the winners are members of the dominant sex. Among various species of spiders, a male that cohabits on a female's web likewise takes her food, and as a result she weighs less than if no male is there. Among savannah lions, females lose much of their prey to males.

Restraint is rare indeed in animal competition over food. Chimpanzees fight over any food that can be monopolized, but the contests are fiercest over meat, producing a fracas that can often be heard more than a kilometer (half a mile) away. Within seconds of a successful predation by a low-ranking chimpanzee, a dominant male is liable to snatch the entire carcass from the killer. In a large group, the carcass will be torn apart by screaming males desperate for a share. Meat-eating can continue for hours. Those without meat, or with only a small piece, beg hard with upturned hands and reaching mouths. The harder they beg, the more meat they get, often by simply tearing it or pulling it away. Possessors try to escape the pressure by turning their backs or climbing to an inaccessible branch. They occasionally charge at their tormentors or flail the carcass at them. Such tactics buy time but are rarely effective. Persistent begging is normally such a nuisance to the possessor that it reduces the rate at which he can eat, and for this reason he sometimes allows others to take a piece of meat. He occasionally even makes an outright donation to a pushy beggar, who immediately leaves with it. Possessing meat can thus be less rewarding than expected from its food value. Meat brings trouble because it takes time to eat.

The most subordinate individuals get little. In the mayhem of carcass division, females rarely end up with a large piece. Overall, females eat much less meat than males, and their low success rate is clearly due to their poor fighting ability. Females with close social relationships to male possessors may get some meat, but in general, meat has less nutritional impact on the lives of female and young chimpanzees than it does on males. Even sexually attractive females cannot expect meat.

If the first cooks were temperamentally like chimpanzees, life would have been absurdly difficult for females or low-status males trying to cook a meal. Cooked food would have been intensely valuable. Even the act of gathering creates value merely by assembling raw foods into a pile. Cooking only increases its attraction. Subordinate individuals cooking their own meals would have been vulnerable to petty theft or worse. If several hungry dominants were present, the weak or unprotected would have lost much or all of the food. Females would have been the losers, just as they are among chimpanzees. There are no indications that human females or their ancestors have ever been prone to forming the kinds of physical fighting alliances with one another that protect bonobo females from being bullied by males.

Consider the possibility that small groups of tough males could search for signs of a campfire as a way to feed themselves. They would be able to descend on an undefended cook and take his or her food at will—after waiting, perhaps, for the cooking to be done. If this ploy were regularly successful, the males could become professional food pirates, which in turn would mean they would not bother to feed themselves or prepare their own food, adding to their desperation to steal it. Male lions come close to doing this, regularly taking whatever meat they want from kills made by females. This scenario suggests that unless cooks somehow established a peaceful environment in which to work, cooking might not have been a viable method of preparing food at all.

Even humans steal readily in various circumstances, so our species is not inherently uncompetitive. The nervous child with a lunch box in the schoolyard knows the problem as well as the anxious late-night stroller with cash in his pocket. People who have the chance to take from members of a different social network have few qualms about doing so. Farmers living next to hunter-gatherers routinely complain of being robbed. Stealing, cheating, and bullying were prevalent among the troubled Ik in the uplands of northern Uganda observed by cultural anthropologist Colin Turnbull, whose book about them, *The Mountain People*, was said by writer Robert Ardrey to record a society without morality. The Ik were a hunting people who had been kept from their traditional hunting grounds. The result was starvation, disease, and mutual exploitation. Turnbull described an almost complete evaporation of their community spirit: "They place the individual good above all else and almost demand that each get away with as much as he can without his fellows knowing." Turnbull's description shows just how savage people can become when social networks break down and life is tough.

Ethnographers sometimes report cases of theft within stable hunter-gatherer communities. Turnbull described how Pepei, an Mbuti Pygmy, had to cook for himself because he was a bachelor with no female kin. As a result, he was often hungry. Several times he was caught stealing small quantities of food from another cooking pot or someone else's hut, mostly from an old woman who had no husband to protect her. His punishment was public ridicule, receiving food fit only for animals, or a thrashing with a thorny branch. Pepei was forgiven after he ended up in tears.

Since hunter-gatherers are often hungry, one might imagine that food theft would be a daily problem. Like other people living in small-scale egalitarian societies, they have no police or any other kind of authority. A hunter-gatherer woman returns to camp in the middle of the day carrying the raw foods she has obtained. She then prepares and cooks them for the evening meal at her own individual fire. Men might return to camp at any time, alone or in a small group. Many of the foods a woman cooks are edible raw, so they could be eaten before, during, or after the cooking process. If a man returns from the bush feeling hungry and has no one to cook for him, he might be tempted to ask a woman for some food—or even simply take it—rather than doing his own cooking. He can also sneak about the camp at any other time, including night.

Yet such tactics are rare. The relaxed atmosphere Lorna Marshall described for the !Kung is due to a system that keeps the peace at mealtimes among hunter-gatherers and other small-scale societies. The system consists of strong cultural norms. Married women must provide food to their husbands, and they must cook it themselves, though other family members may help. Social anthropologists Jane Collier and Michelle Rosaldo surveyed small-scale societies worldwide. "In all cases," they found, "a woman is obliged to provide daily food for her family." That is why married men can count on an evening meal. As a result, they have little reason to take food from women who are not their wives.

The obligation of wives to cook for their husbands occurs regardless of how much other work each of them do, or how much food they give each other. Sometimes men produce much more than women, as among traditional Inuit of the high Arctic, where the almost wholly animal diet of sea mammals, caribou, and fish was produced entirely by men. A man would hunt all day and would come home to a dinner his wife cooked. Cooking was slow over a seal-oil lamp, and women often had to spend much of the afternoon on the task. Sometimes the whole family went hunting together, but the wife had to return early to have everything ready when her husband and others returned to camp. Even when the time of her husband's return was uncertain, she risked punishment if there was no food available for him. But at least a wife's obligation to cook for a husband was matched by his providing all the food.

On the other hand, in some societies women brought home almost all the food. This happened among the Tiwi hunter-gatherers of northern Australia, a polygynous people who lived in households of up to twenty wives and one man. Women foraged for long hours and still returned in the evening to cook the one meal of the day. There were few animals to hunt. Men mostly contributed occasional small animals, such as goanna lizards, and brought in such little food that they needed women's food production for their own welfare. As one Tiwi husband said, "If I had only one or two wives, I would starve." Men relied on their wives not only for their own food but also to feed others. The possession of surplus food was the most concrete symbol of a Tiwi man's success, allowing him to host feasts and promote his political agenda. Women's high food contribution did not sway the balance of power in their marriages. Despite their economic independence and key role in their husbands' status, they were "as frequently and as brutally beaten by their husbands as wives in any other savage society."

Among the Inuit, Tiwi, and all other small-scale societies on record, fairness in distributing labor among women and men was not the issue. Whether or not wives wanted to do so, they cooked for their husbands. As a result, married men were guaranteed adequate food whether they returned late, tired, and hungry from a day's hunting or came home relaxed and early from discussing politics with a neighbor. The man might have eaten in a courteous manner and have had a friendly or even loving interaction with his wife, but the formal structure of their eating relationship was that he could count on her labor and take a large portion of her food—typically, it seems, the best part.

Peace in the camp is further cemented by the principle that unless a husband gave his blessing, a wife could feed no other man except her close kin. This rule applied to cooked food around the campfire, as well as to the raw food she gathered. Other than her kin and husband, no one else had any right to ask for a share, so she could trudge back to camp secure in the knowledge that she would be able to cook all the food she had obtained. In Western society, we take the principle of ownership for granted. But among hunter-gatherers, this manifestation of private ownership is noteworthy because it lies in remarkable contrast to the obligatory sharing of men's foods in particular, and more generally to a strong ethos of communitywide cooperation.

So however hard a man labors to produce food, in hunter-gatherer societies his rights to the food are a matter of communal decision. A man follows the rules, even if that means he gets nothing from his labor. Sometimes he must allow others to distribute his meat. A common requirement among Native American hunters was for boys making their first kill to carry their prize back to camp and stand by while others cooked and ate it. The practice symbolized the subordination of men to the demands of the group. More often, he divided his food himself. The community might allow him to make personal choices about who to give meat to, but not necessarily. In the western desert of Australia, every large hunted animal had to be prepared in a rigidly defined fashion when it was brought to camp. The hunter's own share of a kangaroo was the neck, head, and backbone, while his parents-in-law received a hind leg, and old men ate the tail and innards. The contrast with women's ownership of their foods is striking. Although women forage in small groups and might help one another find good trees or digging areas, their foods belong to them. The sex difference suggests that the cultural rules that specify how women's and men's foods are to be shared are adapted to the society's need to regulate competition specifically over food. The rules were not merely the result of a general moral attitude.

A woman's right to ownership protects her from supplicants of both sexes. In Australia's western desert, a hungry aborigine woman can sit amicably by a cook's fire, but she will not receive any food unless she can justify it by invoking a specific kinship role. It is even more difficult for a man. A bachelor or married man who approaches someone else's wife in search of food would be in flagrant breach of convention and an immediate cause of gossip, just as a woman would be if she gave him any food. The norm is so strong that a wife's presence at a meal can protect even a husband from being approached. Among Mbuti Pygmies, if a family is eating together by their hearth, they will be undisturbed. But when a man is eating alone, he is likely to attract his friends, who will expect to share his food.

Under this system, an unmarried woman who offers food to a man is effectively flirting, if not offering betrothal. Male anthropologists have to be aware of this to avoid embarrassment in such societies. Cofeeding is often the only marriage ceremony, such that if an unmarried pair are seen eating together, they are henceforward regarded as married. In New Guinea, Bonerif hunter-gatherers rely on the sago palm tree for their staple food year-round. If a woman prepares her own sago meal and gives it to a man, she is considered wed to him. The interaction is public, so others take the opportunity to tease the new couple with jokes equating food and sex, such as, "If you get a lot of sago you are going to be a happy man." The association is so ingrained that a man's penis is symbolized by the sago fork with which he eats his meal. If a man takes his sago fork out of his hair and shows it to a woman, they both know he is inviting her for sex. In that society, for a woman to even look at a man's feeding implement is to break the rule against her constrained food-sharing.

Because interactions occur in public, a husband's presence is not necessary to maintain customary principles. The husband's role is important not so much for his physical presence, but because he represents a reliable conduit to the support of the community. If a wife reported to her husband that another man had inappropriately asked her for food, the accused would be obliged to defend himself to both the husband and the community at large.

This may explain one of the reasons why marriage is important to a woman in these societies. Among the Bonerif, as among many hunter-gatherers, sexual intercourse is not tightly restricted to marriage. Wives are free to have sexual relations with several men at the same time, and may do so even when their husbands protest. Furthermore, they get little food from their husbands. But marriage means that her children will be accepted, according to anthropologist Gottfried Oosterwal. In addition, marriage gives a woman access to the only ultimate authority, which is the set of communal decisions reached by men in the men's house. These decisions represent the "crystallized view of everyone about everything" and are accepted as the right view by the whole community. Having a husband means that when social conflict arises, a good wife has an advocate who is a member of the ultimate source of social control.

A link to the communal authority is critical, because the ability of victims to deter a bully or a persistent pest depends on their being a legitimate member of the community. Hunter-gatherers deal with braggarts, thieves, and violators of other social norms in a consistent way, according to anthropologist Christopher Boehm. They use communal sanctions. Whispers, rumors, and gossip evolve into public criticism or ridicule directed at the accused. If the offender continues to incur public anger, he or she will be severely punished or even killed. The killing is done by one or a few men but will be approved by all the elders. Capital punishment provides the sanction that most completely enforces hunter-gatherer adherence to social norms, and it is in men's hands. Thus by virtue of being married (or, if unmarried, by virtue of being a daughter), a woman is socially protected from losing any of her food. Having a husband or father who is a legitimate member of the group, she is effectively protected by him.

In theory, cultural norms that oblige a woman to feed her husband but no other men could have arisen from a societal goal other than to protect women's foods. Such norms might have arisen from a desire to avoid conflicts in general, or from a concern for reducing adultery in particular. But these alternative explanations are unconvincing because men needed their wives specifically to cook for them, rather than merely to behave in a way that promoted communal civility in general. Cross-cultural evidence described above shows that women's cooking for the family is a universal pattern. From ethnographic reports it seems that this domestic service is often the most important contribution a wife makes to their partnership.

We have already seen that among the Tiwi a man depended on being fed by his wives, and it turns out that the Tiwi case is typical. Hunter-gatherer men suffered if they had no wives or female relatives to provide cooked meals. "An aborigine of this Colony without a female partner is a poor dejected being," wrote G. Robinson about the Tasmanians in 1846. When an Australian aboriginal wife deserts her husband, wrote Phyllis Kaberry, he can easily replace her role as a sexual partner but he suffers because he has lost someone attending to his hearth. The loss is important because a bachelor is a sorry creature in subsistence societies, particularly if he has no close kin. As Thomas Gregor explained for the Mehinaku hunter-gardeners of Brazil, an unmarried man "cannot provide the bread and porridge that is the spirit's food and a chief's hospitality. . . . To his friends, he is an object of pity." Colin Turnbull explained precisely why bachelors among Mbuti Pygmies were unhappy: "A woman is more than a mere producer of wealth; she is an essential partner in the economy. Without a wife a man cannot hunt; he has no hearth; he has nobody to build his house, gather fruits and vegetables and cook for him." Examples like these are so widespread that according to Jane Collier and Michelle Rosaldo, in small-scale societies all men have a "strictly economic need for a wife and hearth." Men need their personal cooks because the guarantee of an evening meal frees them to spend the day doing what they want, and allows them to entertain other men. They can find opportunities for sexual interactions more easily than they can find a food provider.

In societies with no restaurants or supermarkets, the need for a wife can lead a man to desperate measures. Among the Inuit, where a woman contributed no food calories, her cooking and production of warm, dry hunting clothes were vital: a man cannot both hunt and cook. The pressure could drive widowers or bachelors to neighboring territories in an attempt to steal a woman, even if it meant killing her husband. The problem was so pervasive that the threat of stealing women dominated relationships among Inuit strangers:

unfamiliar men would normally be killed even before questions were asked. Lust was not the motivation for stealing wives. "The vital importance of a wife to perform domestic services provided the most usual motive for abduction," according to ethnographer David Riches. Oosterwal recorded a comparable reason for wife stealing in New Guinea, where a woman's domestic contribution was critical because of the sago meal she prepared. Men wanted to give feasts as large as possible, so they needed women to organize the food. This led them to conduct raids on neighboring groups to kidnap wives for sago production. Captured women were put to work at once. Their sexual services were an added bonus.

Another version of the same formula applied to many Tiwi marriages. In this highly polygynous culture, old men took most of the young wives, so more than 90 percent of men's first marriages were to widows much older than themselves, sometimes as old as sixty. The old wives might have been past child-bearing age and physically unattractive, but young men delighted in the marriages because they were then fed. Among one nearby group, the Groote Eylandt Aborigines, adult bachelors were given a teenage boy to do the domestic chores. The teenager was called a boy slave, suggesting that wives may have been similarly perceived as fulfilling a slavelike role.

Although the Inuit and Tiwi offer extreme examples of how hunter-gatherer men acquired wives, the importance of marriage for a man in small-scale societies was universal. Collier and Rosaldo explained that a married man has status because once he has a wife, he need never ask for cooked food and he can invite others to his hearth. He is also likely to eat well because men typically eat before their wives and have the choice of the best food. In Michael Symons's words, men "demand selfless generosity from women." To favor the married man even further, small-scale societies have food taboos such that married men are allowed to eat more of the choice kinds of food than are bachelors or women. Women in these societies often dislike marriage specifically because as wives they are obliged to produce food for men, and they have to work harder than they would as unmarried women.

Inequitable as marriage is in certain respects for hunter-gatherer women, that women have to cook for men empowers them. "Her economic skill is not only a weapon for subsistence, but also a means of enforcing good treatment and justice," wrote Phyllis Kaberry of Australian aboriginal women. A wife who cooks badly might be beaten, shouted at, chased, or have her possessions broken, but she can respond to abuse by refusing to cook or threatening to leave. Such disputes seem to be characteristic mostly of new marriages. Most couples easily develop a comfortable predictability, with wives doing their best to provide husbands their cooked meals and husbands appreciating the effort. Hunter-gatherer women are therefore not normally treated badly, and many ethnographers have concluded that, in comparison to most societies, married women lead lives of high status and considerable autonomy.

Catherine Perlès was right in saying that cooking ends individual self-sufficiency. Cooking need not be a social activity, but a woman needs a man to guard her food, and she needs the community to back him up. A man relies on a woman to feed him, and on other men to respect his relationship with her. Without a social network defining, supporting, and enforcing social norms, cooking would lead to chaos.

It is impossible to know how rapidly cooking would have ended individual self-sufficiency after it was first practiced, but in theory the protective pair-bond system could have evolved quickly. Admittedly, the first cooks were not modern hunter-gatherers, and we know too little about their way of life to confidently judge the effects of cooking on social organization. We do not know how linguistically skilled our ancestors were when cooking was adopted. Language is needed nowadays to enforce culturally understood rules, and because a woman's food is made secure by her being able to report on a thief's activity. But at least we can say that three of the key behavioral elements found in the hunter-gatherer system—male food guards, female food suppliers, and respect for other's possessions—are found in other animals, suggesting that a primitive version of the modern food-protection regime could have evolved rapidly among early cooks.

Gibbons illustrate the role of males as food guards. Pairs of these small tree-dwelling apes defend territories against their neighbors. When pairs meet at a tree in the border zone, males fight hard with each other, and the female of the winning male tends to eat better. While food guards are relatively common in animals, there is only one species in which females have been seen provisioning males: a tiny Australian insect called the Zeus bug. Male Zeus bugs are smaller than females and ride on the backs of their mates like jockeys. Females secrete a waxlike material on their backs that is eaten by the male and has no known purpose except to feed him. Males that have been prevented from eating the female's secretions turn competitive: they steal the female's fresh prey. The researchers who discovered this strange relationship hypothesize that females do better by feeding their riding males than by losing prey to them, perhaps because the waxy stuff contains nutrients that the females do not need. This system has apparently evolved to stop males from interfering with the female's feeding. In other words, females feed males to reward them for behaving well. That is close to the system found in humans.

Male "respect for possession" is found more widely than female provisioning. Competition for mates among desert-living hamadryas baboons from around the Red Sea provides a striking example. Male hamadryas who do not know each other fight intensely over females, but among familiars a male is completely inhibited from interfering with an existing bond. Zoologist Hans Kummer demonstrated this with experiments in which he captured two wild males who came from the same group. He found out which of the males was dominant by putting food between them. He then kept the males in separate cages. While the dominant male was allowed to watch, Kummer introduced an unfamiliar female into the cage of the subordinate. The dominant saw everything, but being in a different cage, he could do nothing to stop the subordinate from interacting with the new female. Inside the pairing cage, the subordinate male approached the female and quickly mated with her. A few minutes later she showed him her approval by grooming him, and by that time a bond was formed.

At this point Kummer introduced the dominant male into the cage where the subordinate male and his new female were enjoying their honeymoon. An hour earlier the dominant had been so superior that he had taken food from his subordinate at will, but now the dominant lost all interest in competing for the female. The dominant showed complete respect for the subordinate's possession of the female. Films of these experiments show the dominant looking anywhere but at the subordinate. The dominant develops an intense fascination with a pebble at his feet, which he rolls and twiddles with a pointed finger. He stares at the clouds as if entranced by the weather. The one direction he does not look is toward the most obvious thing in the cage: the two so recently paired baboons. When paired in equivalent circumstances with an unfamiliar male, by contrast, the dominant baboon shows no such respect. Kummer's experiment identified male bonding as the source of respect between males.

The food guarding, provisioning by females, and respect for possession found in animals are associated with males competing over sexual access to females, but only in humans have they led to households. Something about humans is different from other species. A woman's need to have her food supply protected is unique among primates and provides a sensible explanation for the sexual division of labor.

The proposal that the human household originated in competition over food presents a challenge to conventional thinking because it holds economics as primary and sexual relations as secondary. Anthropologists often see marriage as an exchange in which women get resources and men get a guarantee of paternity. In that view, sex is the basis of our mating system; economic considerations are an add-on. But in support of the primary importance of food in determining mating arrangements, in animal species the mating system is adapted to the feeding system, rather than the other way around. A female chimpanzee needs the support of all the males in her community to aid her in defending a large feeding territory, so she does not bond with any particular male. A female gorilla, however, has no need for a defended food territory, so she is free to become a mate for a specific male. Many such examples suggest that the mating system is constrained by the way species are socially adapted to their food supply. The feeding system is not adapted to the mating arrangement. The consequences of a man's economic dependence takes different forms in different societies, but recall that according to Jane Collier and Michelle Rosaldo, his needing a wife to provide food is universal among hunter-gatherers. Food, it seems, routinely drives a man's marriage decision more than the need for a sexual partner.

Furthermore, food relationships appear to be more tightly regulated than sexual relationships. Among the Bonerif, husbands disapproved of their wives having sex with bachelors, but the bachelors did it anyway. Husbands were relatively tolerant of their wives having sex with other husbands, perhaps because promiscuous sex involved less threat of losing her economic services than did promiscuous feeding. As in many other hunter-gatherer communities, Bonerif attitudes toward premarital sex are particularly open-minded. One girl had sex with every unmarried male in the community except her brother. But when a woman feeds a man, she is immediately recognized as being married to him. Western society is not alone in thinking that the way to a man's heart is through his stomach.

Marriage in the United States affects women and men in different ways. Women tend to work longer hours after marriage, thanks to putting in extra time on household tasks, but men do no more household work than before they marry. The pattern is much the same as Jane Collier and Michelle

Rosaldo found in small-scale societies, where marriage "binds specific people together in a particular, hierarchical system of obligations, requiring that women provide services for husbands."

In Victorian England, the aesthetic writer John Ruskin argued that household labor was divided harmoniously and that women were superior to men. He credited women with greater organizational skills than men and explained that women were therefore better at managing households. But to philosopher John Stuart Mill, it was obvious that women were treated unfairly. Ruskin's gallantry, he said, was "an empty compliment . . . since there is no other situation in life in which it is the established order, and considered quite natural and suitable, that the better should obey the worse. If this piece of talk is good for anything, it is only as an admission by men, of the corrupting influence of power."

Mill's accusation that Victorian British men used power to their own advantage might be applied equally well to all nonindustrial societies. The women living on Vanatinai had as much control over their lives as in any society. They were not regarded as inferior to men, and in the public realm they were not subject to male authority. But even when they were tired and men were relaxing, they still had to cook. Maria Lepowsky does not report what would have happened if a woman had refused to cook, but among hunter-gatherers who are similarly egalitarian, husbands are liable to beat wives if the evening meal is merely late or poorly cooked. When there is a conflict, most women have no choice: they have to cook, because cultural rules, ultimately enforced by men for their own benefit, demand it.

The idea that cooking led to our pair-bonds suggests a worldwide irony. Cooking brought huge nutritional benefits. But for women, the adoption of cooking has also led to a major increase in their vulnerability to male authority. Men were the greater beneficiaries. Cooking freed women's time and fed their children, but it also trapped women into a newly subservient role enforced by male-dominated culture. Cooking created and perpetuated a novel system of male cultural superiority. It is not a pretty picture.

CHAPTER 8

The Cook's Journey

"A great flame follows a little spark."

—DANTE, *The Divine Comedy*

When Jean Anthelme Brillat-Savarin wrote, "Tell me what you eat and I shall tell you what you are," he could not have known how right he was. Even nowadays no one knows how deeply the effects of cooking and the control of fire have been burned into our DNA.

Take the pace of our lives. Compared with great apes, we live a few decades longer and reach sexual maturity more slowly. Our long life spans suggest that our ancestors were good at escaping predators. Across species, those who can escape predators more easily tend to live longer. Tortoises, safe in their shells, have lives measured in decades, far longer than most animals their size. Flying species, such as birds or bats, live longer than those confined to the ground, such as mice or shrews. Even in captivity, terrestrial rodents rarely live more than two years, whereas bats of the same size can live for twenty years. Likewise, gliding animals live longer than their nongliding relatives. Bowhead whales stay so far north that killer whales cannot reach them, and they live more than a hundred years. The longevity of early humans is unknown, but their relative safety during evolution must have owed much to the use of fire to deter predators.

Or consider weaning. Cooked food, being soft, enables mothers to wean their young early. During human evolution, early weaning would have allowed a mother to recover her body condition rapidly after birth, promoting a short interval between births. In addition, the higher energy value of cooked food should have promoted a faster rate of growth for the young. The expected early weaning made possible by a human mother's giving cooked food to her infant would have affected social behavior too. Mothers who weaned their babies early would have larger families than before, an infant and a toddler side by side. The advantages of help given by grandmothers and other kin would have increased. Chimpanzee grandmothers occasionally express interest in their daughters' offspring through carrying or grooming, but they are normally preoccupied with their own infants. By generating easily donated gifts of cooked food that are useful for the young, the new system of processing food would have favored cooperative tendencies in rearing families.

Cooking also should have reduced the difficulties of finding enough to eat during the poorest seasons, when even now hunter-gatherers routinely find conditions hard. The notion of cooked food making life easier challenges the thrifty-gene hypothesis, which claims that because the environments of our hunter-gatherer ancestors were highly seasonal, we are physiologically adapted to periods of feast and famine. Accordingly, ancestral humans supposedly digested and stored energy in their bodies with exceptional efficiency. The thrifty-gene hypothesis suggests this efficiency was a useful adaptation when starvation was a consistent threat but is responsible for obesity and diabetes in many modern environments. The cooking hypothesis suggests a different idea: during our evolution, our use of cooked food would have left us better protected from food shortages than the great apes are, or than our noncooking ancestors were. It implies that humans easily become obese as a result of eating exceptionally high-energy, calorie-dense food, rather than from ancient adaptation to seasonality. Great apes become obese in captivity on a rich diet of cooked food.

Cooking and the control of fire must have had substantial influences on our ancestors' digestive physiology. Compared with our close ape relatives, humans regularly experience a higher caloric intake in a short time (e.g., a rapidly ingested evening meal), a more easily digested protein intake, and a higher concentration of the dangerous Maillard compounds that are produced by the combination of sugars and amino acids during cooking. We can therefore expect to find changes in our insulin system compared with those of apes, in the nature of our proteolytic enzymes, and in our systems of defense against a range of carcinogens and inflammatory agents. We might find that we are better protected against Maillard molecules than other primates are, given our uniquely long exposure to ingesting them in high concentrations.

Anthropologists often propose that when fire was first controlled, one of its major contributions was to keep people warm, but that idea wrongly implies that our precooking ancestors would have had difficulty staying warm without fire. Chimpanzees survive nights exposed to long, cold rainstorms. Gorillas sleep uncovered in high, cool mountains. Every species other than humans can maintain adequate body heat without fire. When our ancestors first controlled fire, they would not have needed it for warmth, though fire would have saved them some energy in maintaining body temperature.

But the opportunity to be warmed by fire created new options. Humans are exceptional runners, far better than any other primate at running long distances, and arguably better even than wolves and horses. The problem for most mammals is that they easily become overheated when they run. After a chimpanzee has performed a five-minute charging display, he sits exhausted, panting and visibly hot, beads of sweat glistening among his erect hairs as he uses increased air circulation and sweat production to dissipate his excessive heat. Most mammals cannot evolve a solution to this problem, because they need to retain an insulation system, such as a thick coat of hair. The insulation is needed to maintain body heat during rest or sleep, and of course it cannot be removed after exercise. At best it can be modified, such as by hair being erected to promote air flow.

The best adaptation to losing heat is not to have such an effective insulation system in the first place. As physiologist Peter Wheeler has long argued, this may be why humans are "naked apes": a reduction in hair would have allowed *Homo erectus* to avoid becoming overheated on the hot savanna. But *Homo erectus* could have lost their hair only if they had an alternative system for maintaining body heat at night. Fire offers that system. Once our ancestors controlled fire, they could keep warm even when they were inactive. The benefit would have been high: by losing their hair, humans would have been better able to travel long distances during hot periods, when most animals are inactive. They could then run for long distances in pursuit of prey or to reach carcasses quickly. By allowing body hair to be lost, the control of fire allowed extended periods of running to evolve, and made humans better able to hunt or steal meat from other predators.

The hair loss that benefited adults would have been a problem for babies because babies spend a lot of time inactive and are therefore at risk of becoming cold unless cuddled or nestled in warm surroundings. Perhaps at first babies retained their body hair even when their older siblings lost theirs. But an infant lying next to a fire would have risked burning his or her body hair. Nowadays, human babies are unique among primate infants in having an especially thick layer of fat close to the skin. Baby fat could well be partly a thermal adaptation to the loss of chimpanzee-like hair.

Even our ancestors' emotions are likely to have been influenced by a cooked diet. Clustering around a fire to eat and sleep would have required our ancestors to stay close to one another. To avoid lost tempers flaring into disruptive fights, the proximity would have demanded considerable tolerance. The first dogs provide a provocative model for how tolerance might have evolved. According to biologists Raymond and Lorna Coppinger, wolves began their evolution into dogs when they were drawn to human villages in search of food refuse about fifteen thousand years ago. The Coppingers suggest that when wolves were attracted to these potent new food resources, there was intense natural selection in favor of the calmer individuals, because the calmer wolves were able to get closer to the settlements and more easily find the precious new foods. In effect, dogs experienced a form of self-domestication.

The first cooks probably experienced a similar process. Among the eaters of cooked food who were attracted to a fireside meal, the calmer individuals would have more comfortably accepted others' presence and would have been less likely to irritate their companions. They would have been chased away less often, would have had more access to cooked food, and would have passed on more genes to succeeding generations than the wild-eyed and intemperate bullies who disturbed the peace to the point that they were ostracized by a coalition of the calm. A version of this system had probably already started before cooking, when groups of habilines clustered about a meat carcass.

A process similar to domestication could then have led to an evolutionary advance in ancestral humans' social skills. In animals, more tolerant individuals cooperate and communicate better. Among chimpanzees, individuals that are more tolerant of each other cooperate better. Again, bonobos are more tolerant than chimpanzees, and they collaborate more readily to obtain food. Experimentally domesticated foxes are likewise more tolerant than their wild ancestors and are better at reading human signals. If the intense attractions of a cooking fire selected for individuals who were more tolerant of one another, an accompanying result should have been a rise in their ability to stay calm as they looked at one another, so they could better assess, understand, and trust one another. Thus the temperamental journey toward relaxed face-to-face communication should have taken an important step forward with *Homo erectus*. As tolerance and communication ability increased, individuals would have become better at reaching a mutual understanding, forming alliances, and excluding the intolerant. Such changes in social temperament would have contributed to a growing ability to communicate, including the evolution of language.

The changes wrought by cooked food would have included family dynamics and their supporting psychological mechanisms. The development of pair-bonds in early humans (or their elaboration, if habilines had already evolved a pair-bonding system) contributed to the importance of romantic attachments. On the other hand, domestic violence would have been promoted by the way in which, thanks to cooking, labor is sexually divided and exchanged. Hunter-gatherers are not the only cultures in which wife-beating can be stimulated by disappointments over cooking. Sociologist Marjorie DeVault studied American households and found that "expectations of men's entitlement to service from women are powerful in most families, [and] that these expectations often thwart attempts to construct truly equitable relationships and sometimes lead to violence." Sigmund Freud thought the control of fire led to self-control. Around a hearth, he said, we have to suppress a primal urge to quench the flames with a stream of urine. Freud's notion is far-fetched, but he was right about one thing: our species must have changed radically when we learned to live with flames.

The changes all depend on the mysterious initial moment. We may never know for sure how cooking started, because the breakthrough happened so long ago and probably rather quickly in a small geographical area. But we can use our growing knowledge of great ape behavior, nutrition, and archaeology to speculate. Consider first the woodland apes, or australopithecines. By the period between three million and two million years ago, several genera and many species of australopithecines had already occupied the African woodlands for perhaps three million years. At that time, the only known species of australopithecines were *Australopithecus afarensis*, *A. garhi*, and *A. africanus*, and then even they disappeared.

Climate change appears responsible for the extinction of australopithecine species. Africa began getting drier about three million years ago, making the woodlands a harsher and less productive place to live. Desertification would have reduced the wetlands where australopithecines would have found underwater roots, such as cattails and water lilies, and they would have found fewer fruits and seeds. The species of *Australopithecus* had to adapt their diet or go extinct. Two lines survived.

One adapted by intensifying its reliance on the underground foods that had provided the backup diet of less preferred foods for australopithecines in times of food scarcity. Their descendants rapidly developed enormous jaws and chewing teeth, and are recognized in the naming of a new genus, *Paranthropus*, or the "robust" australopithecines. *Paranthropus* emerged around three million years ago, possibly descendants of *Australopithecus afarensis* or *A. africanus*. They flourished in some of the same dry woodlands as our human ancestors until a million years ago and still looked like upright-walking chimpanzees. But even more than their *Australopithecus* ancestors, *Paranthropus* relied mainly on a diet of roots and other plant storage organs.

The other line of descendants led to humans, and it began with meat eating. Australopithecines must always have been interested in eating meat when they found fresh kills, just as chimpanzees and almost every other primate are today. They would therefore have readily pirated carcasses from any predator they were willing to confront, such as cheetahs or jackals, both of which had close relatives present in Africa by 2.5 million years ago. Chimpanzees today steal carcasses of young antelope or pigs from baboons. But stealing meat from lions and saber-tooths must normally have been too dangerous for australopithecines. Even lions and hyenas kill each other in competition over food, and australopithecines would have been feeble and slow compared to any of the big carnivores.

Given these challenges, it is unclear how australopithecines obtained access to the meat of antelope and other game animals. Maybe they found new ways to kill, which would have given them a few minutes or more to cut meat off their prey before they were chased away by the arrival of big carnivores. Or perhaps they discovered how to stand up to the dangerous predators without serious risk of being wounded or killed. A bold group of australopithecines might have confronted the predators with simple spears modified from digging sticks that they had used to obtain roots. That technology would not have been a huge advance from the short sticks chimpanzees use to jab at bush babies hidden in tree holes, as happens in Senegal. Or maybe they threw rocks at their opponents, much as chimpanzees now sometimes scare pigs or humans with well-aimed missiles in Gombe, Tanzania. If they threw rocks, they might have noticed that sometimes the rocks smashed on landing and produced flakes that could be used for cutting.

Whatever the technique, by at least 2.6 million years ago, some groups were definitely getting meat from carcasses that previously only big carnivores would have eaten. Over the next few hundred thousand years, impact notches and cut marks on animal bones caused by stone tools attest to habilines spending long enough in the danger zones to be able to slice the meat off dead animals, from turtles to elephants. The result was a new and immensely beneficial food source. Knowing that habilines were able to cut steaks and that chimpanzees often pound nuts with hammerstones, we can be sure that habilines would have had the cognitive ability to batter their meat before they ate it, and they surely would have preferred their meat pounded.

Habilines must have also eaten substantial amounts of plant food. During periods of food shortage, such as the annual dry seasons, meat would have been particularly low in fat, down to 1 to 2 percent. Plant foods would then have become critical. Habilines' chewing teeth were similar in size and shape to those of australopithecines, showing a continuing commitment to the same plant foods, including raw roots and corms during the most difficult seasons, and such items as soft seeds and fruits when they could find them. Probably habilines prepared nuts by smashing them to expose the edible seeds, as chimpanzees do. It is doubtful that habilines could process plant foods by any techniques that were much more elaborate than pounding. Almost all the methods hunter-gatherers use to improve the nutritional value of plant foods involve fire, because heat is needed to gelatinize starch. Until fire was controlled, habilines would have been stuck with eating raw plant foods whose caloric value could not be much improved by cold processing.

The breakthrough could have been simple, because it did not require that fire be made from scratch. If fire could be captured, the tending would have been relatively easy. Among hunter-gatherers, children as young as two years old make their own fires by taking sticks from their mothers' fires. Even chimpanzees and bonobos can tend fires well. The bonobo Kanzi is famous for his ability to communicate with psychologist Sue Savage-Rumbaugh using symbols. During an outing in the woods, Kanzi once touched the symbols for "marshmallow" and "fire." He was given matches and marshmallows, and he proceeded to snap twigs for a fire, light them with matches, and toast the marshmallows. By the time of habilines, brain size had roughly doubled compared with the relative brain size of great apes. It is very likely that habilines were mentally capable of keeping a fire alive.

The big question for the habilines that became *Homo erectus* is not how they tended fire, but how they would regularly have obtained it. In his *Descent of Man,* Charles Darwin mentioned an idea suggested by his archaeologist friend John Lubbock: sparks produced by accident from pounded rocks could have launched the control of fire. Anthropologist James Frazer liked the idea of human fire coming accidentally from hitting rocks together, and so did the Yakuts of Siberia, whose campfire tales recounted how hammering led to controlled fire. Certainly habilines would have seen sparks when they hit stones together to make tools. If they softened their meat by pounding it not only with logs but also with hammerstones, they would have had a second source of sparks. There often would have been dry tinder close by, such as grass or the tinder fungus that many people use today to catch a fire.

Anthropologists caution that the sparks produced by many kinds of rock are too cool or short-lived to catch fire. But when pyrites, a common ore containing iron and sulfur, are hit against flint, the result is a set of such excellent sparks that pyrites and flint are standard components of fire-making kits used by hunter-gatherers from the Arctic to Tierra del Fuego. If a particular group of habilines lived in an area exceptionally rich in pyrites, they could have found themselves inadvertently making fire rather often.

The steps to managing fire need not have involved the difficult process of deliberately making it. Here is an alternative scenario: during the tens of thousands of generations between the origin of habilines (at least 2.3 million years ago) and *Homo erectus* (at least 1.8 million years ago), from time to time the sparks resulting from habilines' pounding rocks could have accidentally produced small fires in adjacent brush. Perhaps cocky juvenile habilines dared to grab the cool end of a branch and tease one another with the smoldering twigs or blazing leaves, much as young chimpanzees playfully bully one another with sticks they use as clubs. Adults learned the effect on one another of waving a burning log. The practice of scaring others with fire was then transferred to the serious job of frightening lions, sabertooths and hyenas, similar to how chimpanzees use clubs against leopards. At first the fires went out. But over time, when sparks happened to start a fire, habilines learned that it was worth their while to keep it going. They cultivated fire as a way to help them defend against dangerous animals.

There are other possibilities. The climate was become increasingly dry. Natural fires could have become more frequent. Perhaps people walked behind brush fires looking for cooked seeds. Maybe they obtained fire from trees that burned slowly after being struck by lightning; a eucalyptus tree can smolder for eight months. Perhaps there was a permanent natural source somewhere in Africa, like the gas-fired strip of flame that has been burning nonstop near Antalya in southwestern Turkey ever since Homer recorded it in the Iliad almost three thousand years ago.

Repeated experience with natural fire would have been necessary to give individuals the confidence to use it, which would not have happened easily—otherwise, fire would have been controlled by every group of habilines. But if there were a natural source of fire, such as sparks, there would have been no need to learn to make fire, because it could be taken from nature again and again, and eventually from other groups: the chance of a rainstorm extinguishing every fire in a neighborhood would soon have become vanishingly small. Among Australian aborigines, groups that lost their fire from a drenching rain or flood would refresh their supply from neighbors, who would expect something in return, such as quartz flakes or red ocher. Sometimes the trade occurred across a territorial boundary, which made it dangerous, but risk did not prevent the vital recovery of fire.

Keeping a fire lit would have been a big achievement, but logs are easy to keep aflame when people are moving. Hunter-gatherers regularly carry fire in the form of a burning log. As long as the carrier is walking, the fire is well oxygenated and the log continues to smolder. When people stop, they start a small fire within a few minutes by adding a few sticks to the smoldering log and blowing.

An important step in fire's becoming a central part of human lives was to maintain it at night. Suppose some habilines carried a smoldering log by day to protect against predators, then left it at the base of a sleeping tree when they climbed to make a nest for the night. It would not have been such a big step to give it extra fuel so the log would still be burning the next day—perhaps after seeing this happen first by accident. From there it would have been a smaller step to sitting near the fire to keep it burning, and thereby take advantage of its protection, light, and warmth.

Once they kept fire alive at night, a group of habilines in a particular place occasionally dropped food morsels by accident, ate them after they had been heated, and learned that they tasted better. Repeating their habit, this group would have swiftly evolved into the first *Homo erectus*. The newly delicious cooked diet led to their evolving smaller guts, bigger brains, bigger bodies, and reduced body hair; more running; more hunting; longer lives; calmer temperaments; and a new emphasis on bonding between females and males. The softness of their cooked plant foods selected for smaller teeth, the protection fire provided at night enabled them to sleep on the ground and lose their climbing ability, and females likely began cooking for males, whose time was increasingly free to search for more meat and honey. While other habilines elsewhere in Africa continued for several hundred thousand years to eat their food raw, one lucky group became *Homo erectus*—and humanity began.

EPILOGUE

The Well-Informed Cook

Cooking launched a dietary commitment that today drives an industry. The popular foods cooked in giant factories are often scorned as lacking in micronutrients, having too much fat, salt, and sugar, and having too few interesting tastes, but they are the foods we have evolved to want. The result is excess. By the turn of the twenty-first century, 61 percent of Americans were "overweight enough to begin experiencing health problems as a direct result." With the ready availability of such products as high-fructose corn syrup, cheap palm oil, and intensely milled flour, measured daily energy intake in the United States rose by almost two hundred calories between 1977 and 1995. As a result, more people continue to die in the United States of too much food than of too little, as John Kenneth Galbraith first noted a half century ago. The trends toward easier foods and greater obesity are now found in many industrialized countries. To reverse the decline in health, we should eat more foods with a low caloric density. But few examples can be found in the typical supermarket, because we tend not to like them. We would find it easier to choose appropriate foods if we had a better sense of how many calories we obtain from them. We need to become more aware of the calorie-raising consequences of a highly processed diet.

To do so, we need to better understand nutritional biophysics. Consider meat: the biochemistry of protein digestion is well known. Researchers know precisely what secretions are applied to food molecules at each point in their journey down the alimentary canal. They can say which chemical bonds are severed by which enzymes at which point, how the cells and membranes carry the products of digestion across the gut wall, and how mucosal cells respond to changing pH or mineral concentrations. The detail of biochemical knowledge is exquisite.

Yet this impressive expertise concerns protein, not meat, digestion. Nutritional science is focused so intensively on chemistry that physical realities are forgotten. Researchers treat the food entering the stomach as if it were a solution of nutrients ready for a cascade of biochemical reactions. They forget that our digestive enzymes interact not with free proteins but with a slimy three-dimensional bolus, which after a meal of meat is a messy collection of chewed chunks of muscle, each piece of which is wrapped in multilayered tubes of connective tissue. Structural complexity matters because it affects how easily the food bolus is converted to digestible nutrients, and therefore how many calories we get from our food. As we saw in chapter 3, the rats that gained an extra 30 percent fat in Oka's experiment had no extra calories in their food. They merely had their diet softened. The Evo Diet, described in chapter 1, was calculated to give the volunteers sufficient calories to maintain weight, yet they lost weight rapidly.

Assessing the energy value of foods is a difficult technical problem. Nutritionists cannot calculate the value of foods directly because foods are too complicated in their composition and structure, and digestive systems treat different foods in different ways. So instead of making precise calculations of exactly the number of calories people can obtain from a given food, nutritionists make rough guesses. They do so according to a set of agreed rules that are not perfect but provide a good approximation, at least for foods that are very easily digested. They call these rules a convention.

For more than a century, the convention that has dominated estimating energy values in foods, and now undergirds the food-labeling system of the Western world, has been the Atwater system. Wilbur Olin Atwater, who invented it, was born in 1844. He was a professor of chemistry at Wesleyan College in Connecticut at the end of the nineteenth century. His admirable aim was to ensure that poor people could use their limited resources to get enough to eat. He set out to discover how many calories different foods provided. Atwater knew that food contained three main items the body uses for energy: protein, fat, and carbohydrate. Using a simple laboratory device called a bomb calorimeter, he recorded how much heat was released when typical proteins, fats, and carbohydrates were completely burned. He found that there was not a lot of variation among the different types of each item. For example, all proteins tended to produce a little more than four kilocalories of heat per gram.

After that, Atwater needed to know two more things. First was how much of the major macronutrients—protein, fat, and carbohydrate—a food contains. Fat was easy, because unlike protein and carbohydrates, fat dissolves in ether. So Atwater chopped foods finely, shook them up with ether, and weighed how much material was dissolved in the ether. That gave him a food's fat content (or, more strictly, lipid content: lipids include both fats, which are solid at room temperature, and oils, which are liquid). The same method is used today. Protein was harder to index because no test identifies proteins in general. However, Atwater knew that about 16 percent of the weight of an average protein was nitrogen. So he found a way to measure the amount of nitrogen, which gave him the concentration of protein.

Carbohydrates were the hardest. There was no test then, nor is there now, for identifying the concentration of carbohydrates in general. But Atwater knew the main organic matter in foods were the three big items, protein, fat, and carbohydrate. He also knew how to calculate the total amount of organic matter. He simply burned the food completely, leaving only the mineral ash that did not burn and was therefore the inorganic part. Knowing how much organic matter the food contained and how much fat and protein it held, he obtained the amount of carbohydrate by subtraction: the weight of carbohydrate was what was left when the weights of the fat, protein, and mineral ash had been subtracted from the total weight of the original food item.

Atwater was thus able to estimate the amount of protein, lipid, and carbohydrate in his food. The second piece of information he needed was how much of the food a person eats is digested, as opposed to being passed through the body unused. This required him to analyze the feces of people who were eating precisely measured diets, which he duly did. He was then able to estimate, for each of the three nutrients, how much of what was eaten was also digested. Once again he found that there was little variation within the categories of protein, fat, and carbohydrate, so he assumed the variation could be ignored.

The chemist now had what he wanted. He knew how much energy each of the three big types of macronutrient contained, how much of each macronutrient was present in the food, and how much of it was used in the body. Ignoring variation within each type of macronutrient, he proposed the convention that still dominates the food industry and government standards. By taking into account the proportion of the food that he found was not digested, which was rarely more than 10 percent, he claimed that on average proteins and carbohydrates each yield four kcal/gram, while lipids yield nine kcal/gram. These are known as Atwater's general factors.

This simple and convenient system forms the basis of the Atwater convention and is essentially what the U.S. Department of Agriculture's National Nutrient Database and McCance and Widdowson's *The Composition of Foods* use to produce their tables of nutrient composition. But nutritionists have long recognized important limitations in the Atwater convention, so it has been modified in various ways. One way was to make the general factors more specific. In 1955 the Atwater specific-factor system was introduced to take advantage of a half century of nutritional biochemistry research. For example, the energy value of different types of protein is known to vary: egg protein produces 4.36 kcal/gram, whereas brown rice protein produces 3.41 kcal/gram, and so on. An exhaustive list of such variants has been compiled.

Modifications have also been made to the systems for analyzing nutrient composition. Atwater assumed that all the nitrogen in a food was part of a protein and that all proteins contained 16 percent by weight of nitrogen. However, nitrogen can be found in other molecules that may or may not be digestible, such as nonprotein amino acids and nucleic acids, and some proteins have more or less than 16 percent of nitrogen. So for several decades Atwater's general average of 16 percent nitrogen per protein has been replaced by specific figures, such as 17.54 percent for macaroni protein and 15.67 percent for milk protein.

I mention these modifications to the Atwater system to show that nutritionists have been actively engaged in trying to improve it, and to show that the changes they have proposed have on the whole been rather minor. For example, although egg protein produces more kilocalories per gram (4.36) than brown rice protein (3.41), neither figure is very far from Atwater's estimate of 4 kcal/gram. In fact, although the specific-factor system lends greater precision, the overall effects of the changes are so small that some nutritionists (particularly those in Britain) still prefer to use general factors, albeit modified since Atwater's time.

The general factors have never been static; more factors have been added over time. Even Atwater modified his own system by separating alcohol into its own category (he gave it a rounded value of seven kcal/gram). Much later, in 1970, a general factor was added for the class of carbohydrates called monosaccharides, or simple sugars. New general factors have also been proposed for dietary fiber (or nonstarch polysaccharides), which are so much less well digested than other carbohydrates that they clearly deserve a lower energy value than four kcal/gram; a figure of two kcal/gram has been proposed. The system has also been modified to allow for energy lost in urine and gas production. These and similar modifications continue to tweak the original Atwater system while retaining its essential philosophy.

The Atwater system is thus a flexible convention that is continuously modified but still provides the fundamental basis for assessment of energy value in today's foods. It allows people eating ordinary cooked foods to track their caloric intake sufficiently to get a good idea of when they are overeating or undereating. But it has two critical problems that undermine its ability to assess the food value of items of low digestibility, such as raw foods or foods like whole-grain flour with large particles.

The first problem is that the Atwater convention does not recognize that digestion is a costly process. When we eat, our metabolic rate rises, the maximum increase averaging 25 percent. The corresponding figures for fish (136 percent) and for snakes (687 percent) are vastly higher, showing that humans pay less for digestion than other species, presumably due partly to our food being cooked. But the cost of digestion is still significant for humans and can be reduced or raised depending on the food type.

When Atwater burned foods in a bomb calorimeter, he ignored this complexity. He assumed that humans could use all the energy present in a food and digested in the body. If food burns in the bomb calorimeter, Atwater seemed to conclude, it produces the same amount of energy value in our bodies. But the human body is not a bomb calorimeter. We do not ignite food inside our bodies. We digest it, and we use calories to pay for this complex series of operations. The cost varies by nutrient. Protein costs more to digest than carbohydrates, while fat has the lowest digestive cost of all macronutrients. In a 1987 study, people eating a high-fat diet achieved the same weight gain as others eating almost five times the number of calories in the form of carbohydrate.

The higher the proportion of protein in the food, the higher the cost of digestion. Based on animal studies, we can expect that the costs of digestion are higher for tougher or harder foods than softer foods; for foods with larger rather than smaller particles; for food eaten in single large meals rather than in several small meals; and for food eaten cold rather than hot. Individuals vary too. Lean people tend to have higher costs of digestion than obese people. Whether obesity leads to a low cost of digestion or results from it is unknown. Either way, the variation is important for someone watching his or her weight. For the same number of measured calories, an obese person, having a lower digestive cost, will put on more pounds than a lean person. Life can be unfair.

Compounding the problem, a second big failure of the Atwater system is closely related and equally important. The Atwater system assumes that the proportion of food digested is always the same, regardless of whether the food is in liquid or solid form, part of a high-fiber or low-fiber diet, or raw or cooked. Recall that one of Atwater's general factors was the proportion of food that is passed into the feces undigested. He found that this was low—10 percent or less—and he assumed that this proportion was constant. This assumption has long been known to be wrong. When A. L. Merrill and B. K. Watt introduced the Atwater specific-factor system in 1955, they noted specifically that the digestibility of a grain is affected by how finely it is milled. More extensively milled flour might be completely digested, whereas less milling could lead to 30 percent of the flour being excreted unused. So they called for specific data to be applied to the digestibility of every food. Such data, however, are often unavailable. Identifying the digestibility of each food according to its physical state is difficult, because large numbers of experiments are required. To complicate matters further, the digestibility of the same item varies according to the food context in which it is consumed. For example, protein digestibility tends to be lower when the protein is part of a high-fiber food than when part of a low-fiber food. For raw foods, we have only scattered information on how various durations of cooking, down to none at all, influence the proportion of a food that is digested. Very few studies use the only appropriate measure, ileal digestibility, which takes the sample of unused food at the end of the small intestine, rather than at elimination from the body.

All these factors play such an important role in determining the net value of a food item that many nutritionists have called for a major revision of the Atwater convention. But the information needed to account for the effects of variation in the cost of digestion and digestibility is hard to obtain and difficult to incorporate into a food-labeling system. A widespread preference therefore persists among professionals to keep the Atwater general-factor system. Essentially, nutrition science is faced with choosing between the immense effort of accumulating nutritional-value data that are difficult to quantify but accurate, on the one hand, or using easily quantified but physiologically unrealistic measures, yielding only a rough approximation of food value.

Given the difficulty of acquiring the actual, contextually adjusted nutritional values of individual foods (and combination of foods), the general public is provided with estimates of food values that do not reflect the realities of the digestive process. The scientists who compiled the National Nutrient Database and *The Composition of Foods* must have known that raw foods would produce less net energy than cooked foods and that a higher proportion of raw food was likely to pass through the body unused. But they were locked into an old, approximate-measurement technique, and the result is a falsehood. The data in standard nutritional tables assume that particle size does not matter and that cooking does nothing to increase the energy value of foods, when abundant evidence shows the opposite to be true.

The physics of food matters because our foods and food processing techniques are changing in ways we can expect to contribute to the obesity crisis—thanks to our inability to assess the real caloric value of our diet. In our grocery stores, we find flour that has been ground ever finer, foods made ever softer, calories in ever greater concentration. Rough breads have given way to Twinkies, apples to apple juice. Consumers are misled by the current food-labeling system into thinking they will get the same number of calories from a given weight of macronutrients regardless of how it has been prepared. People are unlikely to experience consequences of our dietary choices any differently than the snakes that got more food value from eating meat that had been ground up, or the rats that got fat when their food pellets were soft. Only one study has been conducted to test the effect of food hardness on health. It found that Japanese women whose diets were softer had larger waists, which are associated with higher rates of mortality. That was a preliminary study. It will take time to show how consistent such effects are, but the indications are clear. We become fat from eating food that is easy to digest. Calories alone do not tell us what we need to know.

It is time to modify Atwater's convention to include the effects of the physical structure of foods in estimates of a food's nutritional value. And we must educate ourselves. As food writer Michael Pollan has argued, we should choose "real food," not "nutrients." For Pollan, real food is natural or only lightly processed, recognizable and familiar. By contrast, nutrients are invisible chemicals, such as essential oils and amino-acids and vitamins, objects of scientific expertise whose significance we must take on faith. The less processed our food, the less intense we can expect the obesity crisis to be.

We once thought of our species as infinitely adaptable, particularly in our diet. Different peoples survive on diets that range from 100 percent plants to 100 percent animals. Such flexibility buttresses a notion that human evolutionary success depended merely on inventiveness. Taken to extremes, our species seems to be free to create our own evolutionary ecology.

The cooking commitment says otherwise. The human ancestral environment was full of uniform problems: how to get fuel, how to regulate feeding competition, how to organize society around fires. The big problem of diet was once how to get enough cooked food, just as it is still for millions of people around the world. But for those of us lucky enough to live with plenty, the challenge has changed. We must find ways to make our ancient dependence on cooked food healthier.

ACKNOWLEDGMENTS

I am indebted to many sources, friends, and colleagues who have guided my attempts to understand the significance of cooking. I owe special thanks to Rachel Carmody, NancyLou Conklin-Brittain, Jamie Jones, Greg Laden, and David Pilbeam for collaboration in research. I am particularly appreciative of those who gave editorial and scholarly advice on earlier versions. Dale Peterson, the late Harry Foster, Martin Muller, Elizabeth Ross, and Bill Frucht commented in generous detail. Rachel Carmody, Felipe Fernandez-Armesto, Elizabeth Marshall Thomas, Victoria Ling, Anne McGuire, David Pilbeam, and Bill Zimmerman also kindly read entire drafts. For comments on individual chapters, I thank Robert Hinde, Kevin Hunt, Geoffrey Livesey, Bill McGrew, Shannon Novak, Lars Rodseth, Kate Ross, Stephen Secor, Melissa Emery Thompson, and Brian Wood. For other kinds of support, ideas, and advice, I am grateful to Leslie Aiello, Ofer Bar-Yosef, Dusha Bateson, Pat Bateson, Joyce Benenson, Jennifer Brand-Miller, Alan Briggs, Michelle Brown, Terry Burnham, Eudald Carbonell, John Coleman, Matthew Collins, Randy Collura, Debby Cox, Meg Crofoot, Roman

Devivo, Irven DeVore, Nancy DeVore, Nate Dominy, Katie Duncan, Peter Ellison, Rob Foley, Scott Fruhan, Dan Gilbert, Luke Glowacki, Naama Goren-Inbar, John Gowlett, Peter Gray, Barbara Haber, Karen Hardy, Brian Hare, Jack Harris, Marc Hauser, Kristen Hawkes, Sarah Hlubik, Carole Hooven, Sarah Hrdy, Stephen Hugh-Jones, Kevin Hunt, Dom Johnson, Doug Jones, Sonya Kahlenberg, Ted Kawecki, Meike Köhler, Kat Koops, Marta Lahr, Mark Leighton, Dan Lieberman, Susan Lipson, Julia Lloyd, Peter Lucas, Meg Lynch, Zarin Machanda, Bob Martin, Chase Masters, the late Ernst Mayr, Rob McCarthy, Rose McDermott, Eric Miller, Christina Mulligan, Osbjorn Pearson, Alexander Pullen, Steven Pyne, Eric Rayman, Philip Rightmire, Neil Roach, Diane Rosenberg, Lorna Rosen, Norm Rosen, Kate Ross, Stephen Secor, Diana Sherry, Riley Sinder, Catherine Smith, Barb Smuts, Antje Spors, Michael Steiper, Nina Strohminger, Michael Symons, Mike Wilson, Tory Wobber, Brian Wood, and Kate Wrangham-Briggs. For exceptional collegial support, I acknowledge the late Jeremy Knowles, Doug Melton, and David Pilbeam. For calm writing opportunities, I thank the staff of the Weston Public Library (Massachusetts), Alison and Kenneth Ross (Badachro, Scotland), Robert Foley and Marta Lahr (Leverhulme Center for Human Evolutionary Studies, Cambridge, UK), the Medical Library of the University of Cambridge (UK), and the authorities of Kibale National Park, Uganda, where I wrote the proposal for this book during three weeks under a fig tree in April 2001.

Acknowledgments

My interest in cooking comes largely from trying to understand reasons for similarities and differences between the behavior of chimpanzees and humans. I have been fortunate to have had the opportunity to study chimpanzee behavioral ecology in Kibale National Park, Uganda, and Gombe National Park, Tanzania. For the financial support that made the Kibale studies possible, I am grateful to the National Science Foundation, Leakey Foundation, National Geographic Society, MacArthur Foundation, and Getty Foundation. For collaboration, I thank especially Adam Arcadi, Colin Chapman, Kim Duffy, Alexander Georgiev, Ian Gilby, Jane Goodall, David Hamburg, Kevin Hunt, Gil Isabirye-Basuta, Sonya Kahlenberg, John Kasenene, Martin Muller, Emily Otali, Amy Pokempner, Herman Pontzer, Anne Pusey, Melissa Emery Thompson, and Michael Wilson.

The late Harry Foster took a gamble when he supported this book, and I much regret that he did not live to see it finished. The support of Amanda Moon, Elizabeth Stein, and Bill Frucht at Basic Books, and the patience of John Brockman and Katinka Matson, were critical.

This project has been immensely rewarding but it intruded grievously into my family's lives. With apology and love, this book is for Ross, David, and Ian, and most especially for Elizabeth.

NOTES

Introduction: The Cooking Hypothesis

2 **The fossil record:** For human evolution, see Klein (1999), Wolpoff (1999), Lewin and Foley (2004). Popular: Zimmer (2005), Wade (2007), Sawyer et al. (2007).

3 **sharp flakes dug from Ethiopian rock:** Toth and Schick (2006).

5 **some habilines evolved into *Homo erectus*:** The fossils I refer to as habilines are conventionally called either *Australopithecus habilis* or *Homo habilis:* Haeusler and McHenry (2004), Wood and Collard (1999). I call them habilines because they do not fit tidily into either *Australopithecus* or *Homo*. The dates of origin and disappearance of the habilines and *Homo erectus* are not precisely known. The most recent evidence of a habiline is at 1.44 million years ago (Koobi Fora, Kenya, specimen number KNM-ER 42703, Spoor et al. [2007]), while *Homo erectus* is possibly seen as early as 1.9 million years ago (KNM-ER 2598), and definitely by 1.78 million years ago (KNM-ER 3733, Antón [2003]). This means *Homo erectus* could have overlapped with habilines for almost half a million years, though the two species did not necessarily occupy the same areas at the same times. Characteristics of *Homo erectus*: Aiello and Wells (2002), Antón (2003).

5 **some anthropologists call them *Homo sapiens*:** Antón (2003, p. 127) reviews the naming debate.

6 **According to the most popular view:** Cartmill (1993) reviews the history of meat-eating hypotheses. Recent advocates for the importance of meat eating in human evolution and adaptation include Stanford (1999), Kaplan et al. (2000), Stanford and Bunn (2001), and Bramble and Lieberman (2004). O'Connell et al. (2002) provide a critique.

10 **"probably the greatest [discovery], excepting language, ever made by man":** Darwin (1871 [2006]), p. 855. Accounts of learning to make fire, and reports of camping days that ended with a cooked evening meal, are in Darwin (1888).

10 **He cited his fellow evolutionist Alfred Russel Wallace:** Darwin (1871 [2006]), p. 867.

12 **"People do not have to cook their food":** Lévi-Strauss (1969); Leach (1970), p. 92.

12 **The celebrated French gastronomist Jean Anthelme Brillat-Savarin:** Brillat-Savarin (1971), p. 279.

12 **ideas suggesting how the control of fire:** Coon (1962), Brace (1995), Perlès (1999), Goudsblom (1992). Quotes are from Symons (1998), pp. 213, 223; Fernandez-Armesto (2001), p. 4.

14 **Those claims constitute the cooking hypothesis**: Wrangham et al. (1999), Wrangham (2006). Collard and Wood (1999) and Wood and Strait (2004) briefly argued in favor of cooking as a stimulus for the evolution of *H. erectus*.

One: Quest for Raw-Foodists

15 **Mongol warriors of the thirteenth century**: Polo (1926), p. 94.

16 **nine volunteers:** The Evo Diet experiment was described by Fullerton-Smith (2007).

17 **Raw-foodists are dedicated**: Many contemporary devotees insist on their diets being 100 percent raw, but most of those who call themselves raw-foodists are not so strict, in some cases allowing half of their diet to be cooked. Most

raw-foodists are vegans, eating diets of germinated seeds, sprouts, cereals, nuts, vegetables, and fruits. Oils and oily fruits such as avocado can be particularly important (Hobbs [2005]).

17 **There are only three studies of body weight in raw-foodists:** Koebnick et al. (1999), Donaldson (2001), Fontana et al. (2005). The Koebnick study has the largest sample and widest range of diets, but all had similar results. Donaldson (2001) studied vegetarians. On a diet of dehydrated barley juice and nineteen daily servings of fruit and vegetables, the subjects felt better than when they had eaten cooked food, but their energy intake was 20 percent below recommended levels. Women took in a mere 1,460 calories per day and men 1,830 calories per day. Fontana et al. (2005) studied raw-foodists and controls matched by age and height. Women who ate raw food weighed 12.6 kilograms less (27.7 pounds) than their counterparts who ate cooked food, while the equivalent drop for men was 17.5 kilograms (38.5 pounds).

18 **no difference in body weight between vegetarians and meat eaters:** Rosell et al. (2005).

18 **"I'm almost always hungry":** Journalist Jodi Mardesich's diary was posted at *www.slate.com/id/2090570/entry/2090637/.*

19 **The Giessen Raw Food study found that 82 percent:** Koebnick et al. (2005).

19 **healthy women on cooked diets rarely fail to menstruate:** Barr (1999), who also reported that among women with stable body weight, vegetarians had fewer menstrual disturbances than those who ate meat.

20 **Reduced reproductive function means:** Ellison (2001) describes the impact of activity on reproductive function.

21 **Anthropologist Elizabeth Marshall Thomas describes:** Thomas (1959).

22 **nutritional biochemist NancyLou Conklin-Brittain finds that carrots contain:** Conklin-Brittain et al. (2002).

22 **Anthropologist George Silberbauer reported:** Silberbauer (1981).

23 **periodic shortages of energy like this are routine:** Jenike (2001).

23 ***Self Healing Power!*** Fry et al. (2003).

23 **They report a sense of well-being:** Hobbs (2005), Donaldson (2001).

23 **reductions in rheumatoid arthritis and fibromyalgia symptoms:** Hobbs (2005).

23 **"Natural nutrition is raw":** Arlin et al. (1996).

23 **Many follow the pseudoscientific ideas of vegetarian Edward Howell:** Howell (1994).

24 **Other raw-foodists are guided by moral principles:** Symons (1998), p. 98, cites Greek sources on the unnaturalness of cooking and meat eating.

24 **the poet Percy Bysshe Shelley:** His argument was published privately in 1813 as *A Vindication of Natural Diet.* Shelley's wife, Mary Shelley, was so inspired by her husband's ideas about the corrosive influences of cooking that when she wrote *Frankenstein* in 1818, she subtitled it *The Modern Prometheus* (Shelley [1982]). Like imagined ancestors in a golden age, Frankenstein's created human (the "monster") was a vegetarian who at first ate his food raw: he relied on finding berries in trees or lying on the ground. When Frankenstein's monster found a campfire abandoned by beggars, he discovered that cooking improved the taste of offal. Mary Shelley thus echoed old ideas that the importance of cooking lay in better tastes. However, she appeared to acknowledge that humans now need cooked food, because the monster declared himself to be similar to real humans in almost all ways except that he could survive on a coarser diet. She herself ate her food cooked.

25 **Instinctotherapists:** Devivo and Spors (2003).

26 **Recent studies indicate that low bone mass:** Fontana et al. (2005). Other health consequences: Koebnick et al. (2005).

26 **What about reliance on raw food in nonindustrialized cultures?** Sumerians: Symons (1998), p. 256. "Only savages" is

from the Chevalier Louis de Jaucourt, cited by Symons (1998), p. 100. Seri: Fontana (2000), p. 22. Fontana (2000), p. xxvii, said much of what McGee wrote about the Seri has been thoroughly discredited. McGee wanted to prove the Seri would be primitive, and made unsupported claims to buttress his preconceptions. Felger and Moser (1985), p. 86, describe Seri cooking. They wrote that the "numerous earlier accounts of the Seri eating raw or even spoiled meat may be somewhat exaggerated or secondhand information." Pygmies in Ruwenzoris: *New Vision* (Uganda newspaper), March 2, 2007, citing the executive director of Uganda's Rural Welfare Improvement for Development. Pygmies have been much studied everywhere. Pygmies cook their food, from Cameroon to Uganda. There have been many parallel claims about the existence of tribes who do not know how to make fire. Again, these claims have been carefully examined and found to be wrong. Particular individuals may be poor at making fire, however. Furthermore, there can be times when people do not have the relevant tool kit at hand, such as fire stones, drills, and tinder. For universal cooking: Tylor (1870), p. 239. For universal fire-making: Frazer (1930), Gott (2002).

27 **The quirky nutritionist Edward Howell thought so:** Howell (1994).

27 **The most detailed studies of unwesternized Inuit diets were by Vilhjalmur Stefansson:** Stefansson's diaries are detailed in Pálsson (2001), pp. 95, 97, 100, 204, 210, 282. See also Stefansson (1913), pp. 174, and Stefansson (1944).

28 **"Woe betide the wife who keeps him waiting":** Jenness (1922), p. 100.

30 **like every culture the main meal of the day was taken in the evening:** Quote is from Tanaka (1980), p. 30. Evidence of hunter-gatherers eating evening meals: Inuit: "The one cooked meal of the day was in the evening," Burch (1998), p. 44; Tiwi: "at least two or three of [my wives] are likely to bring something back with them at the end of the day, and then we can all eat," Hart and Pilling (1960), p. 35; Aranda: "The principal meal is generally taken toward evening,

when returning from the hunt and the *mana*-gathering. The women collect the fuel," Schulze (1891), p. 233. Siriono: "The principal meal is always taken in the later afternoon or early evening," with each nuclear family cooking its own food, Holmberg (1969), p. 87; Andaman Islanders: "In the afternoon the women return with what food they have obtained and then the men come in with their provision. The camp, unless the hunters have been unsuccessful, is then busy with the preparation of the evening meal, which is the chief meal of the day. . . . The meat is distributed amongst the members of the community and the woman of each family then proceeds to cook the family meal," Radcliffe-Brown (1922), p. 38; Tlingit: "Formerly they ate but twice daily: the morning meal upon rising . . . and evening food." "The latter was the substantial meal . . . the hunter or traveler would not eat until he was safe in camp, or the day's work done," Emmons (1991), p. 140. I have found no exceptions to the evening meal being the main reported meal for hunter-gatherers.

31 **Most fruits are preferred raw and are eaten in the bush:** The proportion of plant species always eaten cooked was markedly higher for roots (76 percent of fifty-one species), seeds (76 percent of forty-five species) and nuts (75 percent of sixteen species) than for fruits (5 percent of ninety-seven species). Data tallied from Appendix in Isaacs (1987). Raw snacks by day: Australians: O'Dea (1991); Peruvians: Johnson (2003).

31 **Dougal Robertson, and his family lost their boat to killer whales:** Robertson (1973).

33 **The case that comes closest to long-term survival on raw wild food is that of Helena Valero:** Valero and Biocca (1970), chapter 13.

34 **Anthropologist Allan Holmberg was at a remote mission station in Bolivia:** Holmberg (1969), p. 72.

35 **Robert Burke and William Wills led an ill-fated expedition:** Murgatroyd (2002).

35 **it is rare for people to even attempt to survive on raw food in the wild:** Pacific: Heyerdahl (1996); Andes: Read (1974); Essex: Philbrick (2000); Japanese: Onoda (1974).

36 **"he got fire by rubbing two sticks of Piemento Wood together upon his knee":** Quote is from Woodes Rogers, in Letterman (2003), p. 73.

Two: The Cook's Body

38 **Spontaneous benefits are experienced by almost any species:** Effects of cooked food on farm animals: Mabjeesh et al. (2000), Campling (1991), Pattanaik et al. (2000), Medel et al. (2002), Medel et al. (2004), Nagalakshmi et al. (2003). In cattle there is a limit to this relationship, because cows need a minimal amount of roughage in their diet (Owen [1991]).

38 **Salmon grow better on a diet of cooked rather than raw fishmeal**: Stead and Laird (2002). Although cooked fishmeal was developed in 1937 and chicken pellets were invented in 1944, the value of cooking has only recently been appreciated by the fish-farming industry. Salmon farming, the most important form of British aquaculture, depends importantly on fishmeal, which provides 20 percent to 35 percent of the worldwide aquaculture foods. The main sources of fishmeal are small oceanic species such as anchovy and sardine. About six million to seven million metric tons of fishmeal were cooked, pressed, dried, and ground annually around the turn of the twenty-first century. In the 1980s, commercial diets in early British salmon farms were cheap because they used a conventional pellet press process without extrusion, in which the temperature of the material rose to a mere 60°C–70°C (140°F–158°F) and pellets were cut off after being pushed through a press like bits of pasta. Salmon prices in Britain were high, around £7 per pound, so salmon-farm owners made adequate profits even though the fish grew relatively slowly and fewer survived compared

to nowadays. Then the price of salmon began to fall, putting economic pressure on the farmers and making the right food choice more important. Fish-feed manufacturers started to use intense cooking methods, producing extruded feeds. Fishmeal and grain ingredients were pressurized with water and superheated steam at temperatures up to 120°C (248°F) before being passed through a die under pressure. The increased heat led to more gelatinization of starch and more effective killing of pathogens. The pellets were also puffed up by "flashing off" of water during the extrusion process, thought to increase digestibility. Although diet may not have been the only contributor to rising success, it accounted for almost half of production costs, so its effectiveness strongly affected profits. With the change in food processing, the industry's performance improved. During the 1990s, the average weight of harvested fish rose from 2.5 kilograms (5.5 pounds) to almost 4 kilograms (8.8 pounds), survival rose from about 60 percent to 90 percent, and production costs fell.

39 **Biologically Appropriate Raw Food, or BARF:** Palmer (2002) discusses raw food for dogs. BARF diets are touted at *www.barfworld.com/html/barf_diet/barfdiet.shtml.*

39 **Even insects appear to get immediate benefits:** Carpenter and Bloem (2002), Fisher and Bruck (2004), Pleau et al. (2002).

40 **humans have an astonishingly tiny opening:** Our mouths are small partly because our lips create small openings compared to those of other primates. The difference is less when bones are compared. Kay et al. (1998) measured oral cavities in forty-eight human and forty-four chimpanzee skulls. They found that human oral cavities were a little smaller (107 cubic millimeters) than those of chimpanzees (113 cubic millimeters). Data presented on thirty-three primates by DeGusta et al. (1999) allow calculation of oral cavity size in arbitrary units, suggesting that humans have marginally larger mouths than chimpanzees, though small in relation to

body weight. Smith and Jungers (1997) summarized body weights. The median wild adult body weight for three subspecies of chimpanzees was 42 kilograms (female) and 46 kilograms (male). For seven human populations ranging from Pygmies to Samoans, the median weights were 53 kilograms (female) and 61.5 kilograms (male). These data indicate that humans weigh 26 percent to 34 percent more than chimpanzees. However, since the measured oral cavities came from European populations, a more realistic estimate of human body weights (from Denmark) is 62 kilograms (female) and 72 kilograms (male). This comparison has humans weighing 48 percent to 57 percent more than chimpanzees. The reason our mouths look particularly small is that they do not project in front of our faces as they do in chimpanzees: our mouths are tucked so much farther back under our skulls that we have more room in them than a look from the outside suggests. Lucas et al. (2006) comment on the effect of cooking on human mouths.

42 **this gene, called MYH16:** Stedman et al. (2004). The detailed work on myosin composition in the jaw muscles is restricted to macaques, but apes are assumed to be similar. Much research remains to be done before the timing of the mutation in the MYH16 gene can be decided with confidence. Recent studies suggest that the mutation may be as old as 5.3 million years. If so, the reasons are puzzling.

42 **Human chewing teeth, or molars, also are small:** Data were kindly given to me by Neil Roach, based on Kay (1975) using teeth of humans twenty-five thousand years old from Predmosti. Soft food leading to small jaws and teeth: reviewed in Lucas (2004), Lieberman et al. (2004). An alternative idea invoking soft foods was proposed by Milton (1993): humans' small teeth could be adapted to soft fruits. But it is normally thought that soft fruits were less available in human diets during the past two million years, when they had small teeth, than in earlier times, thanks to their commitment to terrestriality and savanna habitats.

42 **physical anthropologist Peter Lucas has calculated:** Lucas (2004).

43 **the stomach is less than one-third the size:** Data are from Martin et al. (1985), based on forty primates and seventy-three mammals.

43 **Great apes eat perhaps twice as much by weight per day as we do:** A wild chimpanzee weighing 41 kilograms (90 pounds) eats about 1.4 kilograms (3.1 pounds) of dry weight of food per day (personal observation, Kibale National Park). A Kalahari bushman of the same weight eats a paltry 0.7 kilograms (1.6 pounds)—about half the chimpanzee's intake: urban raw-foodists eat about the same. Relationship between dry weight of daily food intake and body mass of primates and humans: Barton (1992). Modern urban raw-foodists: Wrangham and Conklin-Brittain (2003). Fiber content: Conklin-Brittain et al. (2002).

43 **the human small intestine is only a little smaller than expected:** Martin et al. (1985) show the surface area in humans is smaller than 62 percent of forty-two primate species and is 76 percent of the size expected by comparison with seventy-four mammal species. Milton (1999) notes that our small intestine is long relative to the size of our gastrointestinal system. Though this is true, it has not been shown to be long relative to our body weight. So this does not indicate a special adaptation.

43 **humans have the same basal metabolic rate as other primates:** Leonard and Robertson (1997).

43 **the large intestine, or colon, is less than 60 percent of the mass that would be expected:** Martin et al. (1985) found the surface area of the human colon is smaller than in 92 percent of thirty-eight species of primates in relation to body weight and is 58 percent of the expected size compared to seventy-four mammal species.

43 **The colon is where our intestinal flora ferment plant fiber:** For human reliance on plants, see the consensus noted by Bunn and Stanford (2001), and other chapters in Stanford and Bunn (2001).

44 **the volume of the entire human gut:** Calculated from data in Chivers and Hladik (1980) and Milton and Demment (1988), comparing humans to thirty-five primate species. Gut mass 60 percent of expected: Aiello and Wheeler (1995).

44 **a reduction in the size of jaw muscles:** Lucas et al. (2008) propose that jaw muscles are small in humans because the body needs to be accurate in sensing forces when chewing.

44 **the reduction in human gut size saves humans at least 10 percent:** Aiello and Wheeler (1995), p. 205.

44 **The suite of changes in the human digestive system makes sense:** Wrangham and Conklin-Brittain (2003). Milton (1999), and Stanford and Bunn (2001) reviewed the meat-eating hypothesis.

45 **australopithecines had broad hips and a rib cage:** Aiello and Wheeler (1995).

45 **the molars (chewing teeth) of very early humans were somewhat sharper:** Ungar (2004).

46 **Carnivores such as dogs, and probably wolves and hyenas, also tend to have small guts:** Chivers and Hladik (1980, 1984), Martin et al. (1985), MacLarnon et al. (1986), Milton (1987, 1999). Large guts in australopithecines are indicated by wide flaring of the ribs (Aiello and Wheeler [1995]).

46 **Dogs tend to keep food in the stomach for two to four hours, and cats for five to six hours:** Transit times compared between carnivores and primates: Milton (1999). Transit times compared between humans and dogs using same meals (cooked chicken liver): Meyer et al. (1985, 1988). In humans, 50 percent of the meal was emptied from the stomach after approximately 105 minutes; in dogs, 50 percent of the same meal was emptied after approximately 180 minutes. See also Tanaka et al. (1997), Ragir (2000). Cats: Armbrust et al. (2003).

47 **Raw meat might have been usefully pounded:** The idea that a key behavioral adaptation of early humans was the use of tools to process food goes back at least to Oakley (1962). Milton and Demment (1988) suggested that using

tools could account for reductions in tooth and gut size in the human lineage. Teaford et al. (2002) proposed that reduction in incisor size could likewise be related to increased use of tools for processing food.

47 **It might have been allowed to rot:** Sherman and Billing (2006) discuss the problem of bacterial infection in meat.

48 **"If you are transferred suddenly from a diet normal in fat":** Quote is from Stefansson (1944), p. 234. Stefansson's years of ethnographic work with the Inuit made him intensely interested in their dietary adaptations, and he conducted several intriguing experiments on himself. Speth (1989) describes how Stefansson lived on meat alone for a year in New York under medical supervision. Mostly his diet was 25 percent protein and 75 percent fat, but he adjusted it to an intake of 45 percent to 50 percent protein for a time. He then experienced nausea, diarrhea, loss of appetite, and general discomfort. He felt better again within two days of returning to the diet of 25 percent protein. For maximum levels of protein, see Speth (1989).

49 **People with an anatomy like ours today could not have flourished on raw food:** An alternative idea could be that marrow, which requires no chewing, could have been eaten at sufficiently high levels to promote protein- and fat-digesting specializations in the gut while allowing the mouth, jaws, and teeth to be small. However, although marrow may well have been an important component of the diet, it cannot have been exclusive in view of the high frequency of cut marks on the bones of prey animals around the time these features changed in human evolution.

50 **Take, for example, Maillard compounds:** Vlassara et al. (2002) review health problems associated with these compounds.

51 **The tastes are strong and rich:** Nishida (2000) systematically catalogued the tastes of chimpanzee foods in the Mahale Mountains, Tanzania.

53 **Even when we cook our meat:** Ragir et al. (2000), Sherman and Billing (2006).

Three: The Energy Theory of Cooking

55 **authoritative science flatly challenges this idea:** U.S. Department of Agriculture, *National Nutrient Database for Standard Reference* (2007). McCance and Widdowson's *The Composition of Foods:* Food Standards Agency (2002). To assess the apparent effect of cooking, I compared calorie densities per dry weight for foods where nutrient data were reported both for the raw and cooked versions. In some cases, minor gains in energy were reported, such as a 1.7 percent increase in the energy density of carrots from being boiled, or a 1.5 percent increase in sirloin after roasting. In others there were minor losses in energy density, such as a 1.8 percent decrease in the energy density of beets after boiling, or a 2.0 percent loss in energy of tenderloin after being roasted. Overall, such cases canceled each other out. When I plotted a graph showing the energy density in cooked foods against the energy density in raw foods, I found that on average, cooked foods were reported to have almost exactly the same energy density as raw foods, regardless of whether they were rich in carbohydrates or protein.

56 **"a technological way of externalizing part of the digestive process":** Aiello and Wheeler (1995), p. 210.

56 **"fresh premium breakfast sausages":** made by Shady Brook Farms.

56 **Leading nutritionist David Jenkins:** Jenkins (1988), p. 1156.

57 **Starchy foods are the key ingredient:** McGee (2004) is an excellent source on the science of cooking. Wandsnider (1997) discusses the chemistry of cooking using hunter-gatherer technology.

57 **cereals such as rice and wheat made up 44 percent of the world's food production:** Atkins and Bowler (2001), Table 9.4.

58 **Studies of ileal digestibility show that we use cooked starch very efficiently:** Home-cooked kidney beans: Noah et al. (1998); flaked barley: Livesey et al. (1995); cornflakes, white bread, oats: Englyst and Cummings (1985); bananas: Langkilde et al. (2002), Englyst and Cummings (1986),

Muir et al. (1995); potatoes: Englyst and Cummings (1987); wheat: Muir et al. (1995). For review see Carmody and Wrangham (forthcoming).

59 **The principal way cooking achieves its increased digestibility is by gelatinization:** Eastwood (2003) and Gaman and Sherrington (1996) provide textbook accounts; Olkku and Rha (1978) give a detailed review; Svihus et al. (2005) and Tester et al. (2006) discuss frontline research. For an example of effects of baking on starch (i.e., heating in the absence of water), see Karlsson and Eliasson (2003). Lee et al. (2005) illustrate how increasing degrees of gelatinization lead to increased hydrolysis and increased glucose absorption in rats, an example of incomplete starch digestion in animals.

59 **The granules are . . . too small to be seen with the naked eye:** Despite the granules' small size, people may be able to detect them in their food, because foods containing particles as small as two micrometers in diameter (two thousandths of a millimeter, or 0.08 thousandths of an inch) feel rougher when rubbed against the top of the mouth, or between tongue and lips, than foods without any particles. So people may be able to use "mouth-feel" to detect the presence of starch granules. Engelen et al. (2005a) tested human perception of particle size by adding silica dioxide and polystyrene spheres of known size to custard desserts. Until their research, it was thought that perceptions of food smoothness, slipperiness, and roughness were affected only by lubricative properties of the food, like oiliness. That people perceive food with particles the size of even very small starch granules as being rough suggests that we can detect (and avoid) raw starch through its effect on texture.

60 **the glucose chains are unprotected, and gelatinize:** The chains of glucose come in two types, or molecules. Amylopectin is the "good" one. Amylopectin is a huge molecule made up of as many as two million glucose units linked to each other in a rambling, open, branching framework. Following gelatinization, amylopectin offers easy access to di-

gestive enzymes. So starches that are mostly made up of amylopectin satisfy quickly, providing a highly digestible food with a high glycemic index.

The difficult component of starch is amylose because this molecule is resistant to digestion even after gelatinization. On average, amyloses compose 20 percent to 30 percent of starch granules by weight, but their concentration can vary from zero to 70 percent. Amylose is a small molecule made up of only fifty to five hundred glucose units. The units line up in relatively short unbranched chains that can wrap about themselves, sometimes together with lipids, to form hydrophobic structures resistant to penetration and therefore readily protected from amylases and related enzymes. So amylose-rich starches are a good food for someone trying to lose weight or worried about diabetes. Especially at higher concentrations, their presence is a major reason for starch being resistant to digestion. Brown et al. (2003) showed that cooking makes amylose more digestible, though above 60 percent amylose, even cooking did not completely remove the resistance of starch.

60 **The effect of eating cornstarch:** Collings et al. (1981).

61 **Cooking consistently increases the glycemic index of starchy foods:** Brand-Miller (2006).

61 **Even the effects on proteins are a matter of debate:** Reviewed by Carmody and Wrangham (forthcoming).

62 **"An egg should never be cooked":** Christian and Christian (1904), p. 159.

62 **This kind of argument:** Roach (2004) describes controversies among bodybuilders about the value of raw eggs.

63 **When aborigines on the beaches of Australia's tropical north coast are thirsty:** Isaacs (1987), p. 166.

63 **hunter-gatherers prefer to cook them:** Emu eggs: Basedow (1925), p. 125. Yahgan: Gusinde (1937), p. 319.

63 **a Belgian team of gastroenterologists tested the effects of cooking:** Evenepoel et al. (1998, 1999).

64 **in the large intestine bacteria and protozoa digest the food proteins entirely for their own benefit:** Rutherfurd

and Moughan (1998), p. 909: "Amino acids do not appear to be absorbed to any significant extent by the large intestinal mucosa of large mammals."

64 **they were able to check their results with healthy subjects as well:** After feeding the labeled egg meals to the ileostomy patients, the researchers not only collected ileal effluent every thirty minutes but also took samples of their exhaled breath. They found that the course of digestion (monitored from ileal effluents) was closely correlated with the appearance of stable isotopes in the breath. This taught the researchers that breath tests alone would reveal how well the labeled protein was digested. Breath tests were accordingly used to study egg digestion in healthy volunteers.

65 **Cooking increased the protein value of eggs by around 40 percent:** The discovery that we do not digest raw egg protein nearly as well as cooked egg protein is the first to identify the effects of heat on the digestibility of protein in the human gut. But the evidence that raw eggs are a relatively poor food has been hinted at by other studies. For example, allergy researchers collected breast milk from women who had eaten either raw or cooked eggs for breakfast. They found that the concentration of ovalbumin rose in breast milk after eating eggs, and that the rise was about twice as fast when eggs were cooked as when they were raw. Again, the cooked eggs appeared more digestible. Allergy study: Palmer et al. (2005). The recent data about the effects of cooking on digestibility of eggs were anticipated by at least two groups. Hawk (1919) claimed his research team had evidence that raw egg white is used less completely than cooked egg white. Cohn (1936) showed that rats grew poorly on diets rich in raw egg white compared to those eating cooked egg white. She attributed this partly to the anti-trypsin factor and partly to raw egg proteins being passed more rapidly than cooked egg proteins from the stomach to the small intestine, an effect also found by Evenepoel et al. (1998). Cohn's suggestion that a rapid gastric emptying rate might be responsible for the poor energy

supply from raw eggs is not supported by modern data. First, in recent decades the idea that the stomach was responsible for a large proportion of digestion has given way to the orthodoxy that most digestion occurs in the small intestine. Second, Evenepoel et al. (1998) found no difference in transit time to the ileocecal junction (the half-time averaged 5.3 hours in both cases). This meant that raw eggs spent more time than cooked eggs in the small intestine, where digestive processes are most active, so they should have been better digested than the cooked eggs.

65 **Denaturation occurs when the internal bonds of a protein weaken:** McGee (2004), Wandsnider (1997).

65 **In 1987 researchers chose to study a beef protein:** Davies et al. (1987) studied the degradation of bovine serum albumin by trypsin with and without heating. Proteins were four times more easily digested in an experiment in which they were lightly heated. This suggests that in real life when they have been properly cooked they would be much easier to digest.

65 **Acid is vital in the ordinary process of digestion:** The pH of the empty stomach is usually less than 2. This intense acidity is not always regarded by digestive physiologists as important with respect to denaturation. Johnson (2001) and King (2000) both cite the function of gastric acid as bactericidal and converting pepsinogen to pepsin; neither mentions denaturation. By contrast, Sizer and Whitney (2006), p. 81, report that "Stomach acid works to uncoil protein strands and to activate the stomach's protein-digesting enzyme. Then the enzyme breaks the protein strands into smaller fragments."

66 **Marinades, pickles, and lemon juice . . . can contribute to the denaturing of proteins in meat, poultry, and fish:** Gaman and Sherrington (1996).

66 **Hunter-gatherers have likewise been reported mixing acidic fruits with stored meats:** Tlingit: Emmons (1991), pp. 140, 143; pemmican: Driver (1961), p. 71; Australians: Berndt and Berndt (1988), p. 99.

66 **Animal protein that has been salted and dried, such as fish, is likewise denatured:** Sannaveerappa et al. (2004) found that when Indian milkfish were salted for twenty-four hours, their large muscle proteins were substantially denatured. Sun drying exacerbated the effect.

67 **"A large portion of the side was blown off":** Beaumont (1996), p. ix.

68 **"At 12 o'clock, M., I introduced through the perforation, into the stomach":** Beaumont (1996), p. 125.

68 **"The rugae gently close upon it":** Beaumont (1996), p. 77.

68 **"gradual appearance of innumerable, very fine, lucid specks":** Beaumont (1996), p. 104.

69 **"Vegetable, like animal substances, are more capable of digestion":** Beaumont (1996), p. 47.

69 **"Fibrine and gelatine . . . are affected in the same way":** Beaumont (1996), p. 35.

69 **"Pieces of raw potato":** Beaumont (1996), p. 48.

70 **the world's most expensive sandwich:** BBC News, April 10, 2006, *http://news.bbc.co.uk/go/pr/fr/-/1/hi/england/london/4894952.stm*; *www.wagyu.net/home.html.*

71 **"Of all the attributes of eating quality":** Lawrie (1991), p. 199.

71 **the cook's main goal has always been to soften food:** "Softness" is an elusive quality. Hardness is the force needed to initiate a crack. Toughness measures the force needed to keep it going. Springiness tells how fast a deformed food returns to its original shape. Chewiness is the number of times it has to be chewed to make it fit to be swallowed. All those factors contribute to the ordinary perception of softness, or food that "melts" in the mouth. Others matter too, such as juiciness (the rate at which moisture is released) or greasiness (the difficulty of removing a fatty film coating the mouth). Different cuts of meat vary in each of these aspects, and cooking affects each kind of texture in different ways. Lucas (2004) discusses the physics of food. Ruiz de Huidobro et al. (2005) discuss meat textures.

71 **"The central theme is that cooks assist the bodily machine"**: Symons (1998), p. 94.

71 **"to render mastication easy":** Beeton (1909), p. 108.

72 **"it is so tender that the sinews will fall apart":** Tanaka (1980), pp. 38, 39.

72 **The island-living Yahgan:** Gusinde (1937), p. 325.

72 **a delicacy shared by the Tlingit:** Emmons (1991), p. 141.

73 **Game animals have a few soft parts:** Utes: Pettit (1990), p. 44. Australians: Dawson (1881), p. 17. Inuit (intestines): Jenness (1922), pp. 104, 106. Inuit (kidneys and liver): Jenness (1922), p. 100. Chimpanzees: personal observation. Philbrick (2000) reports the eating of raw liver by marine cannibals. Cannibalism normally involved cooking, however.

73 **fat-tailed sheep:** Fernandez-Armesto (2001), p. 88.

73 **While some foods are naturally tender**: Gaman and Sherrington (1996).

74 **the tensile strength of tendons can be half that of aluminum:** Lawrie (1991), chapter 3.

74 **Three left-handed helices of protein twirl:** Woodhead-Galloway (1980).

75 **good cooking tenderizes every kind of meat:** The most widely used index of meat toughness is the Warner-Bratzler shear force, measured by the effort needed to penetrate meat with a steel blade. Warner-Bratzler measurements tend to match consumers' perception of "hardness," but hardness is only one of several components of consumer preference. So taste panels of consumers who chew meat samples provide the best assessment of texture even though they are time-consuming, expensive, and somewhat variable in their results. For instance, consumer perceptions vary across countries. Warner-Bratzler shear force: Harris and Shorthose (1988), Tornberg (1996). Variation across countries: Lawrie (1991). Meat tenderized by cooking: shrimp: Rao and Lund (1986); octopus, Hurtado et al. (2001); rabbit, Combes et al. (2003); goat, Dzudie et al. (2000); beef, de Huidobro et al. (2005).

75 **Steak tartare requires:** Rombauer and Becker (1975), p. 86.

76 **Brillat-Savarin recorded an enthusiastic testimony:** Hunt (1961), p. 17, citing Brillat-Savarin's *Gastronomy as a Fine Art* (1826).

76 **A team of Japanese scientists:** Oka et al. (2003). The average yield forces of hard and soft pellets were 85.5 newtons and 41.8 newtons, respectively.

78 **Secor and his team have shown repeatedly:** Pythons: Secor (2003); toads: Secor and Faulkner (2002). Overview of the costs of digestion: Secor (2009).

79 **But grinding and cooking changed the costs of digestion:** Boback et al. (2007).

81 **Cooked food is better than raw food because life is mostly concerned with energy:** Female chimpanzees: Thompson et al. (2007), Williams et al. (2002). Energy and human reproduction: Ellison (2001).

Four: When Cooking Began

83 **fire was not regularly used for cooking until the Upper Paleolithic:** Jolly and White (1995).

83 **Others favor much earlier times, half a million years ago:** Aiello and Wheeler (1995), Rowlett (1999), Ragir (2000), Foley (2002).

83 **physical anthropologist Loring Brace:** Brace's specific ideas about the effects of cooking on tooth size reduction have not been widely followed, but Brace has done more to stress the potential importance of cooking than most recent anthropologists, and his interpretation of archaeological evidence as indicating that the control of fire began around a quarter of a million years ago seems to have been the dominant view in recent decades (e.g., James [1989] and commentators on his paper).

84 **Abri Pataud:** Bricker (1995).

84 **Abri Romani:** Pastó et al. (2000).

84 **Vanguard Cave:** Barton et al. (1999). Pullen (2005) and Victoria Ling (personal communication) provide excellent reviews of fire evidence from the Lower Paleolithic onward.

84 **Klasies River Mouth:** Pullen (2005).

84 **Sodmein Cave:** Pullen (2005).

85 **Kalambo Falls:** Clark and Harris (1985).

85 **Hayonim:** Albert et al. (2003).

85 **the older part of the record, going back in time from a quarter of a million years ago, has been improving in quality:** Another rich source of fire evidence is the 400,000-year-old site at Bilzingsleben, where Mania ([1995]; Mania and Mania [2005]) has argued that hearths exist outside of the dwellings and a further hearth is in the center of a circular pavement. The hearths take the form of localized and discrete patches of burning upon the ground.

85 **Beeches Pit:** Gowlett (2006), Preece et al. (2006). Beeches Pit also contains some burnt bones. With respect to the reconstruction of events at the Beeches Pit fire, the distribution of artifacts around the hearth suggests that hominins were undertaking fireside knapping. In particular, the refitting of a series of around thirty flakes, two of which were indeed burnt, provides a direct link between the knapping undertaken by an individual and a fire. Although it is unknown whether the fire provided a focus for social interaction, it is a reasonable suggestion given that several different forms of biface were retrieved from this area (Gowlett et al. [2005]). In 2007, John Gowlett kindly took me to this quiet woodland where the slope from a former living site still angles down toward the site of an ancient pond. I squatted precisely where so long ago, someone appears to have knapped an ill-chosen flint by a fire.

86 **Schöningen:** Thieme (2000, 2005). Originally, four spears were reported (Thieme [1997]), but Thieme (2000) reports "more than half a dozen," without being specific as to exact number. One spear was found next to a horse pelvis (Thieme [1997]). All are carved from spruce (*Picea* sp.) except spear

IV, which is made from pine (*Pinus* sp.). They are carved from individual trees with a dense concentration of growth rings. The trees were felled, debarked, and had the side branches removed. The tips of the spears are worked from the hardest part of the wood at the base of the tree. Spear VI is 2.5 meters in length.

87 **Gesher Benot Ya'aqov:** Goren-Inbar et al. (2004).

87 **"had a profound knowledge of fire-making":** Alperson-Afil (2008), p. 1733.

87 **Archaeological sites between a million and a million and a half years old:** James (1989).

88 **Others accept the idea that humans controlled fire in the early days of *Homo erectus* as well established:** Rowlett (1999), Boyd and Silk (2002).

88 **the Hadza:** Mallol et al. (2007).

88 **the half-lives of caves average about a quarter of a million years:** John Gowlett and Alfred Latham, personal communication, November 2006. Swartkrans (more than one million years old) is a cave made of dolomite, which resists erosion.

88 **people must have used fire, yet there is no sign of it:** Recent sites without fire despite abundant evidence of fire in contemporaneous sites in the same region include High Cave in Tangier, Bisitun in Iran, Grotte Suard in Charente (Oakley [1963]). Likewise Sergant et al. (2006) report that in the cover sand area of the northwest European Plain, burnt bone, shells, and artifacts have been found on nearly every Mesolithic site (i.e., within the last ten thousand years prior to the introduction of farming), yet there are no structured hearths and the visibility of the "campfire" is extremely poor or in many sites unknown altogether.

88 **mysterious reductions in the frequency of finding evidence of fire:** Victoria Ling (personal communication, 2007).

90 **Anthropologists have sometimes suggested:** Stahl (1989), p. 19, suggests that "use of controlled fire as a source of warmth

may have preceded systematic use of fire in food preparation by thousands or hundreds of thousands of years."

90 **Victoria Wobber and Brian Hare tested chimpanzees and other apes:** Wobber et al. (2008).

90 **Chimpanzees in Senegal do not eat the raw beans of *Afzelia*:** Brewer (1978).

91 **"Koko indicated the 'tastes better' option":** Penny Patterson, personal communication, May 2007.

91 **sensory nerves in the tongue:** Hiiemae and Palmer (1999).

91 **brain cells (neurons) responsive to texture converge with taste neurons:** Kadohisa et al. (2005b).

91 **such factors as grittiness, viscosity, oiliness, and temperature:** Kadohisa et al. (2004), Kadohisa et al. (2005a).

91 **Edmund Rolls found when people ate foods of a particular viscosity:** de Araujo and Rolls (2004) used fMRI to assess neural responses in twelve subjects being given sucrose, vegetable oil, or solutions of carboxymethyl cellulose of known viscosity. Rolls (2005) gives an overview.

92 **Studies of Galapagos finches by Peter and Rosemary Grant:** Galapagos finches, *Geospiza fortis*: Boag and Grant (1981), Grant and Grant (2002). After the intense selection for larger beaks, food became abundant and beak size returned slowly to its smaller original size. Weiner (1994) describes the Grants' research.

93 **In fewer than eight thousand years, mainland boa constrictors:** Boback (2006).

93 **According to evolutionary biologist Stephen Jay Gould:** Gould (2002).

93 **Loring Brace suggested:** Brace (1995). The pattern of decline in the size of chewing teeth is now known to be more complex than Brace suggested (Bermudez de Castro and Nicolas [1995]).

94 **When fruits are scarce, gorillas rely on foliage alone:** The contrast in diet is probably due to food spending a longer time in the gut of gorillas, allowing more opportunity for fermentation of plant fiber and so giving gorillas more

ability to survive on this lower-quality food. See Milton (1999). Wrangham (2006) compares behavior and ecology of chimpanzees and gorillas.

95 **gorillas mature earlier:** Gorilla first birth is around nine years, compared to around fourteen years for chimpanzees; gorilla interbirth interval averages every 3.9 years, compared with chimpanzees every 5.0–6.2 years (Knott [2001]). A leaf diet might allow a predictable food regime sufficient to permit the evolution of rapid rates of growth and reproduction.

95 **The anatomical differences between a cooking and a pre-cooking ancestor:** Wrangham (2006).

96 **largely complete by around two hundred thousand years ago.** Earliest *Homo sapiens*: White et al. (2003).

96 ***Homo heidelbergensis* was merely a more robust form of human than *Homo sapiens*:** Lieberman et al. (2002).

97 ***Homo heidelbergensis* evolved from *Homo erectus*:** Rightmire (1998, 2004). Cranial capacity increased from around 900 cubic centimeters (54.9 cubic inches) in *Homo erectus* to about 1,200 cubic centimeters (73.2 cubic inches) in *Homo heidelbergensis.*

97 **the original change, from habilines to *Homo erectus*:** Anton (2003), McHenry and Coffing (2000). Areas of chewing teeth are totals for the second premolar and the first two molars. They totaled 478 square millimeters (0.74 square inches) in *Australopithecus (Homo) habilis*, compared to 377 square millimeters (0.58 square inches) in early *Homo erectus.*

98 **42 percent increase in cranial capacity:** *Australopithecus (Homo) habilis*: 612 cubic centimeters (37 cubic inches). *Homo erectus*: 871 cubic centimeters (53 cubic inches) (McHenry and Coffing [2000]).

99 **The only nonhuman primate that regularly sleeps on the ground:** Mehlman and Doran (2002).

99 **early Pleistocene periods in Africa were rich in predators:** Werdelin and Lewis (2005).

100 **The famous "Turkana boy":** Walker and Shipman (1996). *Homo erectus* in general: Antón (2003). Comparison to habi-

lines: Haeusler and McHenry (2004), Wood and Collard (1999). I assume that australopithecines and habilines were all sufficiently good climbers to sleep in trees, following Hunt (1991). Although that appears to be the majority view, Ward (2002) is cautious, suggesting that we cannot be sure how well *Australopithecus afarensis* climbed. However, it seems inconceivable that australopithecines slept on the ground.

101 **Modern hunter-gatherers are safer in camp at night:** Kaplan et al. (2000).

102 **natural selection rapidly favored the anatomical changes that facilitated long-distance locomotion:** Haeusler and McHenry (2004) argue that habilines had long legs (as well as an upper body adapted to climbing). There are only two specimens of *habilis* with sufficient postcranial remains to reconstruct leg lengths, so this is still contested ground. If they are right, the problem of where habilines slept is more complicated than my assumption that they slept in trees.

Five: Brain Foods

105 **Blaise Pascal:** Pascal's *Pensées*, 1670.

106 **Richard Alexander argues:** Alexander (1990).

106 **Death rates from these interactions among chimpanzees are similar:** Wrangham et al. (2006).

107 **Species of primates with larger brains are more intelligent:** Deaner et al. (2007).

107 **but they show no overall tendency to have larger ranges:** Dunbar (1998).

107 **Robin Dunbar found:** Shultz and Dunbar (2007).

108 **the crow family have many of the social abilities of primates:** Cnotka et al. (2008).

108 **Bottlenose dolphins form particularly complex and changeable alliances:** Connor (2007).

108 **Spotted hyenas live in large groups:** Carl Zimmer, *New York Times,* March 4, 2008. See also Holekamp et al. (2007).

108 **"cerebral ganglia of extraordinary dimensions":** Darwin (1871 [2006]), p. 859.

108 **the social brain hypothesis:** Dunbar (1998), Byrne and Bates (2007).

109 **In 1995 Leslie Aiello and Peter Wheeler proposed:** Aiello and Wheeler (1995).

110 **genes that are responsible for energy metabolism show increased expression:** Khaitovich et al. (2008).

113 **The idea became known as the expensive tissue hypothesis:** Fish and Lockwood (2003) supported Aiello and Wheeler's proposal by showing that brain size is related to diet quality in primates. Hladik et al. (1999) suggest that other body parts are also reduced in size to compensate for large brains.

113 **An elephant-nosed mormyrid fish:** Kaufman (2006).

113 **Birds . . . grow bigger wing muscles:** Isler and van Schaik (2006). They suggest that in human evolution, cheaper locomotion might likewise have enabled brains to enlarge.

113 **Species with relatively low muscle mass have been found to have:** Leonard et al. (2007).

114 **In actuality, that phase of our evolution occurred in two steps:** In a public talk at Harvard University in 2008, Leslie Aiello said that on the basis of recent evidence, cooking was likely to have accounted for the rise in brain size in *Homo erectus*.

115 **Chimpanzees have a cranial capacity:** Chimpanzee brain data measured by Adolph Schultz (David Pilbeam, personal communication, 2005). Australopithecine brain data: McHenry and Coffing (2000).

115 **The most likely alternatives were starch-filled roots:** Laden and Wrangham (2005), Hernandez-Aguilar et al. (2007), Yeakel et al. (2007).

116 **they have less indigestible fiber:** Conklin-Brittain et al. (2002).

116 **A dietary change from foliage to higher quality roots is thus a plausible explanation:** Aiello and Wheeler (1995) proposed an alternative idea, that the rise in diet quality for australopithecines came from their eating more hard foods such as nuts and seeds. But this is hard to accept because

such foods are invariably seasonal, creating periods of food shortage when some other food type would have been needed. That necessary fallback food would have determined the minimum size of the gut.

116 **the roughly 450 cubic centimeters (27 cubic inches) of australopithecines:** McHenry and Coffing (2000).

119 **Tenderizing meat would have reduced the costs of digestion:** Meat-drying is another speculative processing mechanism habilines could have employed, leading to protein denaturation and improved food quality.

120 ***Homo erectus* brains continued to increase in size after 1.8 million years ago:** Rightmire (2004).

122 **mongongo nuts eaten by !Kung San:** Lee (1979), p. 193.

123 **Various modern behaviors:** McBrearty and Brooks (2000).

123 **The ovens . . . are not recorded in Australia until thirty thousand years ago:** Brace (1995). Cooking in earth ovens: Smith et al. (2001).

123 **among the Aranda of central Australia:** Spencer (1927), p. 19.

125 **people made a glue from ancient birch tar:** Mazza et al. (2006).

125 **Andaman Islanders . . . cooked:** Cooking methods were described by Man (1932).

126 **the Yahgan developed a two-stone griddle:** Gusinde (1937), pp. 318–320.

Six: How Cooking Frees Men

129 **Chimpanzee society differs markedly:** Mitani et al. (2002), Doran and McNeilage (1998).

129 **Sherwood Washburn and Chet Lancaster wrote:** Washburn and Lancaster (1968), p. 23.

131 **The Hadza are modern-day people:** I spent a few nights in a Hadza camp in 1981 with Monique Borgerhoff-Mulder, but this account comes mostly from reports by ethnographers such as Hawkes et al. (1997, 2001a, 2001b), Marlowe (2003), and Brian Wood (personal communication, 2008).

Note that the Hadza are like almost all hunter-gatherers in having had long-term relationships with neighbors who are farmers and pastoralists (Headland and Reid [1989]).

132 **"They did not have pleasurable satisfaction":** Marshall (1998), p. 67.

132 **Anthropologist Phyllis Kaberry:** Kaberry (1939), p. 35.

133 **the sexual division of labor among hunter-gatherers:** Megarry (1995), Bird (1999), and Waguespack (2005) give overviews.

133 **while men hunted sea mammals, women would dive for shellfish:** Steward and Faron (1959).

134 **In the tropical islands of northern Australia, there was so much plant food:** Hart and Pilling (1960).

134 **women always tended to provide the staples:** "In almost all [societies], the items women tend to focus on are commonly acquired, come in smaller sizes, have a relatively low risk of pursuit failure, and are often associated with high processing costs. The resources men prefer generally are more rarely acquired, larger, have higher risk of pursuit failure, and are associated with lower processing costs." Bird (1999), p. 66. Women's food items were so vital as predictable staples that a principal reason for the group to move camp was overexploitation of women's foods (Kelly [1995]).

134 **a kind of bread called damper:** Isaacs (1987) describes its preparation.

134 **"The Aborigines continually craved for meat":** Kaberry (1939), p. 36.

135 **Hunting large game was a predominantly masculine activity:** In a sample of 185 societies, the only activities that were more male-biased were lumbering, metalworking, ore-smelting, and hunting sea mammals (Murdock and Provost [1973], Wood and Eagly [2002]).

135 **Hints of comparable sex differences in food procurement:** Kevin Hunt (personal communication [2005]), compilation of data on forty primate species.

135 **the overwhelming majority of the foods collected and eaten by females and males are the same types:** Perhaps the

most extreme sex difference in primate diets is that male chimpanzees eat more meat than females. But neither sex eats much meat. Both sexes spend the great majority of their time eating fruits, around 50 percent to 70 percent of their time, so the sex difference in meat-eating by chimpanzees is relatively trivial compared to humans. The highest known recorded meat intake averaged at forty grams per day for males, which probably provides less than 2 percent of total calories (Kaplan et al. [2000], Table 3).

136 **each household is a little economy:** Hunter-gatherer men are often quoted as saying to their wives in the morning, like the Inuit studied by Stefansson, "Make sure you have my evening meal ready for me when I get back." There is nothing equivalent in any nonhuman animal. Yanigasako (1979) reviews the distinction between family and household from the perspective of social anthropology. "Family" connotes a set of relationships, especially genealogical; "households" refer to family members who live together and engage in food production and consumption or in sexual reproduction and child-rearing. Panter-Brick (2002) gives an overview.

136 **It used to be thought that women typically produced most of the calories:** Lee and DeVore (1968).

136 **Worldwide across foraging groups, however, men probably supplied the bulk of the food calories:** In nine groups, on average, women produced 34 percent, men 66 percent of calories (Kaplan et al. [2000]).

137 **Emile Durkheim thought:** Durkheim (1933), p. 56: "We are thus led to consider the division of labor in a new light. In this instance, the economic services that it can render are picayune compared to the moral effect that it produces, and its true function is to create in two or more persons a feeling of solidarity."

137 **"fundamental platform of behavior for the genus *Homo*":** Lancaster and Lancaster (1983), pp. 36, 51.

137 **many think the division of labor by sex started much later:** There is an increasing trend in anthropology and

archaeology to think of the sexual division of labor as developing "recently," i.e., as late as the Upper Paleolithic (around forty thousand years ago) (Steele and Shennan [1996], Kuhn and Stiner [2006]). The trend comes from the difficulty of recognizing gender-differentiated activities archaeologically in earlier periods.

138 **"When males hunt and females gather":** Washburn and Lancaster (1968), p. 301. Washburn did not specifically discuss cooking in the context of the sexual division of labor, but his writings imply that he thought cooking developed later.

139 **Chimpanzees in Gombe National Park, Tanzania, spend more than six hours a day chewing:** Wrangham (1977).

139 **the amount of time spent chewing is related to body size:** Clutton-Brock and Harvey (1977) showed that bigger primates spend more time eating. With an enlarged data set that corrected for errors and ensured a uniform definition of eating as chewing, R. Wrangham, Z. Machanda, and R. McCarthy (unpublished) predicted humans on a raw-food diet would need to chew at least 42 percent of the time. The figure for *Homo* is lower than the figure for Gombe chimpanzees (more than 50 percent), even though humans are heavier than chimpanzees, because the prediction uses data from all primates. Great apes tend to fall above the primate line, which is brought down by the smaller-bodied monkeys. The 42 percent figure is thus a conservative estimate.

140 **A few careful studies using direct observation:** Cross-cultural time-allocation data come from studies inspired by Johnson (1975) and published in a series of monographs by the Human Relations Area Files: Ye'kwana, Hames (1993); Quechua, Weil (1993); Newar, Munroe et al. (1997); Mekranoti, Werner (1993); Logoli, Munroe and Munroe (1991); Kipsigi, Mulder et al. (1997); Samoans, Munroe and Munroe (1990b); Black Carib, Munroe and Munroe (1990a); Machiguenga-Camaná, Baksh (1990); Machiguenga-Shimaa, Johnson and Johnson (1988); Yukpa, Paolisso and Sackett (1988); Madurese, Smith (1995). Hofferth and Sand-

berg (2001) give data for American children. The number of minutes spent eating per twenty-four hours (or percent of nonsleep time, which I calculated from their data) were as follows: 9–12 years, 77 minutes (9.8 percent); 6–8 years, 63 minutes (7.5 percent); 3–5 years, 69 minutes (8.4 percent); 0–2 years, 99 minutes (14.4 percent).

141 **Processed plant foods experience similar physical changes to those of meat:** Plant foods: Waldron et al. (2003). Meat: Barham (2000). Foods from domesticated plants are presumably also softer than their wild counterparts.

141 **softness (or hardness) closely predicts:** Measuring chewing rates in 266 people, Engelen et al. (2005b) found a correlation of .95 between number of chewing cycles before swallowing and food hardness.

141 **Wild monkeys spend almost twice as long chewing:** Agetsuma and Nakagawa (1998) showed that Japanese monkeys spent 1.7 times more time feeding where food requirements were higher and food was lower quality.

142 **a chimpanzee mother who consumes 1,800 calories per day:** Pontzer and Wrangham (2004) estimate energetic expenditure at 1,814 calories per day for chimpanzee mothers in Kanyawara (Kibale, Uganda), and 1,558 calories per day for adult males.

142 **around 300 calories per hour:** Assuming that wild male chimpanzees use 1,558 calories per day (Pontzer and Wrangham [2004]) and chew for six hours, they ingest 260 calories per hour.

143 **less than three minutes per day hunting:** time spent per day is calculated from a median 0.13 hunts per day (Watts and Mitani [2002], Figure 9) and the mean hunt duration (17.7 minutes), giving 2.3 minutes. The estimate is higher than it should be because it assumes that all individuals hunted throughout the duration of a group hunt, which is not true. However, it serves to show that chimpanzees spend only a short time hunting per day.

143 **A recent review of eight hunter-gatherer societies:** Waguespack (2005). Hadza men: Hawkes et al. (2001b).

143 **at Ngogo the longest hunt observed:** Watts and Mitani (2002).

143 **the average interval between plant-feeding bouts was twenty minutes:** Data are from Gombe males, 348 inter-feeding intervals in 628 observation hours (1972–1973), median 20.3 minutes, mean 43.5 minutes (Wrangham, unpublished data).

145 **suppose the male has had an unsuccessful day of hunting:** Only around 50 percent of hunts by chimpanzees are successful, and even when a kill is made, there is no guarantee that any particular male will be able to get any meat to eat. Hunting success: Gilby and Wrangham (2007). For the Hadza "records of over 250 camp-days of observation across all seasons over a period of five years show several stretches of a week or more with no meat from big game available" (O'Connell et al. [2002]).

Seven: The Married Cook

148 **Overall cooking was the most female-biased activity:** Women did the cooking "almost exclusively" in 63.6 percent of societies and "predominantly" in 34.2 percent. After cooking, the next most female-biased activities were preparing vegetable food (mostly by women in 94.3 percent of societies), fetching water (91.4 percent), and doing the laundry (87 percent) (Murdock and Provost [1973]).

148 **the Todas:** The idea that Toda men were responsible for cooking was derived from Murdock's misreading of Rivers (1906), who conducted fieldwork among the Todas. Marshall (1873), p. 82 referred to women invariably cooking the daily meals, and Breeks (1873) stated that while the men fetched firewood, the women cooked and fetched water. Prince Peter (1955) conducted his own fieldwork and corrected Murdock's error.

149 **The procedure for cooking the fruit pulp:** Truk: Gladwin and Sarason (1953); Marquesans: Handy (1923).

149 **women were responsible for cooking:** For both women and men, "a large proportion of daily activity . . . is devoted to the production or preparation of food" (Gladwin and Sarason [1953], p. 137). The distinction in breadfruit-eating societies between communal cooking by men and domestic cooking by women is the most extreme example of a system found in many societies. At communal events such as feasts, ritual meals or even just the cooking of a large animal, men tend to be the cooks. On such occasions, as with the cooking of breadfruit, men cook in groups and share the product with one another (Goody [1982], Subias [2002]).

150 **"tremendous overlap":** Lepowsky (1993), p. 290.

150 **"personal autonomy":** Lepowsky (1993), p. xii.

151 **"We come home"**: Lepowsky (1993), p. 289.

151 **The word *lady*:** Hagen (1998).

151 **The classic reason:** For example, according to psychologists Wendy Wood and Alice Eagly, "certain activities are more efficiently accomplished by one sex. It can thus be easier for one sex than the other to perform certain activities of daily life under given conditions. The benefits of this greater efficiency emerge because women and men are allied in complementary relationships in societies and engage in a division of labor" (Wood and Eagly [2002], p. 702). The same kind of explanation has been widespread in evolutionary scenarios. Marlowe (2007) found that in environments where more plant foods were available, men did more gathering. Women tend to obtain foods that are compatible with child care, while men take responsibilities for other tasks (Marlowe [2007]). Becker (1985) reviews evidence that the sexual division of labor is beneficial for household efficiency in the United States.

152 **"the sex-relation is also an economic relation":** Gilman (1966 [1898]), p. 5.

152 **"has made of woman a slave":** Christian and Christian (1904), p. 78.

152 **"The culinary act":** Perlès (1977), translated by Symons (1998, p. 213) .

152 **"if only to ensure":** Goudsblom (1992), p. 20.

153 **Fernandez-Armesto proposed:** Fernandez-Armesto (2001), p. 5.

153 **"the starting-place of trades":** Symons (1998), p. 121. Symons poetically summarized the importance of cooking as an act of sharing by saying that sauces "dispense goodness."

154 **Examples of individual self-sufficiency clearly undermine the idea that the sheer mechanics of cooking require that it be practiced cooperatively:** Archaeologist Martin Jones captures the uncertainties of explaining how cooking and cooperation are related in his 2007 book, *Feast*, subtitled *Why Humans Share Food.* Jones considered that the antecedents of human food-sharing lie in a basic primate tendency, seen in the occasional giving of food by primate mothers to their offspring. Humans built on this proactive generosity, Jones suggests, when our African ancestors responded to a shortage of important plant foods by hunting more. The demands of the hunt led to cooperation, bigger brains, and cooking. "The unique abilities of the modern human brain brought us to a most unusual behavioural pattern, the gathering around a hearth in a conversational circle to share food" (Jones [2007], p. 299). This may be right, but it leaves open many possibilities for exactly how cooking and cooperation were related.

155 **the sight or smell of smoke reveals a cook's location:** Tindale (1974) records Australian aborigines traveling forty kilometers (twenty-five miles) to steal fire.

156 **"When a visitor comes":** Marshall (1998), p. 73.

156 **predictably induce fights:** Competition over meat: Goodall (1986). Importance of food's ability to be monopolized: Wittig and Boesch (2003). Breadfruit: Hohmann and Fruth (2000). Male lions living on open plains often steal food from females (unlike the more widespread male lions that live in woodlands, which mostly hunt for themselves): Funston et al. (1998). Spiders: Arnqvist et al. (2006).

157 **The harder they beg, the more meat they get:** Gilby (2006).

158 **Even sexually attractive females cannot expect meat:** Stanford (1999), p. 212, wrote that a male chimpanzee "withholds a scrap of meat from a female until she mates with him." Similar assertions are cited widely and go back to the 1970s. Detailed analysis now shows that the success of a female in getting meat is not affected by her sexual status and that a female who receives meat does not have an increased probability of mating (Gilby [2006]). Furthermore, the probability of hunting falls when sexually receptive females accompany males (Gilby et al. [2006]). Gilby et al. (2006) suggest the old concept that chimpanzees exhibit "meat-for-sex" needs to be replaced with a new idea: "meat-or-sex."

158 **There are no indications:** Most males of species in the human lineage were not only larger than females, but also exhibit features associated with more aggressive behavior than would be found in females. In particular, there seem to have been important differences between the sexes in the breadth of the face, with males having wider faces characteristic of aggressive behavior. Bonobos are the only great apes in which females are able to defend food from males, even though they are smaller than males. But male bonobos have relatively narrow, young-looking facial masks compared to the more aggressive chimpanzees. Early hominid anatomy shows no evidence of a bonobo-like style of feminized males (Wrangham and Pilbeam [2001]).

159 **complain of being robbed:** Turnbull (1965), Grinker (1994).

159 **"They place the individual good":** Turnbull (1974), p. 28.

160 **Pepei:** Turnbull (1965), p. 198.

161 **"In all cases":** Collier and Rosaldo (1981), p. 283.

161 **traditional Inuit:** Jenness (1922), especially p. 99.

161 **Tiwi . . . of northern Australia:** Hart and Pilling (1960). The full quote is "If I had only one or two wives I would starve, but with my present ten or twelve wives I can send them out in all directions in the morning and at least two or three of them are likely to bring something back with them at the

end of the day, and then we can all eat." Thus women can share food through their relationship to a man. The amount of food produced in the household was critical to a man's prestige: "the most concrete symbol of Tiwi success was the possession of surplus food" (p. 52). Beating quote: p. 55.

162 **typically, it seems, the best part:** Kelly (1993) discusses that food taboos favor men, because the taboos (which prohibit certain classes of people from eating meat) apply more to women than to men. An example of hunter-gatherer men eating better than women, with known health consequences, is given by Pate (2006) for southeastern Australia.

163 **A common requirement among Native American hunters:** Driver (1961), p. 79.

163 **In the western desert of Australia, every large hunted animal:** Hamilton (1987), p. 41.

164 **The rules were not merely the result of a general moral attitude:** Rules for sharing men's foods are reviewed by Kelly (1993).

164 **a hungry aborigine woman:** Hamilton (1987), p. 42.

164 **Mbuti Pygmies:** Turnbull (1965), p. 124. Compare Andaman Islanders: "while, however, all members of a family take their meals together, a married man is only permitted to eat with other (married men) and bachelors, but never with any women save those of his own household, unless indeed he be well advanced in years. Bachelors as well as spinsters are required to take their meals apart with those of their respective sexes." (Man [1932], p. 124).

164 **effectively flirting, if not offering betrothal:** Mbuti: Turnbull (1965), p. 118. Collier and Rosaldo (1981) review hunter-gatherers in which marriages start without ceremony, but merely by living together.

165 **Bonerif hunter-gatherers:** Oosterwal (1961), p. 82, reports on several tribes in the Tor territory, including Bonerif and Berrik in particular. Their patterns are mostly similar, and I have called them the Bonerif here to represent all of them. Oosterwal (1961), p. 95, notes that women could offer sago

to him only through their husbands; otherwise, their action would have been misinterpreted.

166 **according to anthropologist Christopher Boehm:** Boehm (1999).

166 **The killing is done by one or a few men:** For example, Lorna Marshall (1998), p. 84, heard about only one theft of food among the !Kung Bushmen. A man took honey from a honey tree that had been found, marked, and therefore owned by someone else. The furious owner killed him for it. The murder went unpunished, tacitly approved by the group.

167 **the Tasmanians:** Robinson (1846), p. 145.

167 **an Australian aboriginal wife:** Kaberry (1939), p. 36.

168 **"cannot provide the bread":** Gregor (1985), p. 26.

168 **bachelors among Mbuti Pygmies:** Turnbull (1965), p. 206.

168 **"strictly economic need":** Collier and Rosaldo (1981), p. 284. Among the Bonerif, bachelors had so little to eat that they normally left camp and roamed (Oosterwal [1961], p. 77). Among the Bonerif, the men who were best off were the newly married, because their wives were young and strong. Bachelors with no mothers or sisters had little to eat, and men who wanted more food got married even if it meant raiding neighboring groups, at risk of death and subsequent revenge.

169 **"The vital importance of a wife":** Riches (1987), p. 25.

169 **wife stealing in New Guinea:** Oosterwal (1961), p. 117.

169 **many Tiwi marriages:** Hart and Pilling (1960).

169 **boy slave:** Rose (1960), p. 20.

170 **"demand selfless generosity":** Symons (1998), p. 171. Symons stresses that although sharing is in his view the essence of cooking, the sharing is not fair.

170 **"Her economic skill is not only a weapon for subsistence":** Kaberry (1939), p. 36.

170 **A wife who cooks badly might be beaten:** Consequences of bad or late cooking: Mbuti, Turnbull (1965), p. 201; Siriono, Holmberg (1969), p. 127; Inuit, Jenness (1922); Bonerif,

Oosterwal (1961), p. 94. Sulking wife refusing to cook food: Mbuti, Turnbull (1965), p. 276.

172 **Gibbons illustrate:** Fuentes (2000).

172 **Zeus bug:** Arnqvist et al. (2006).

172 **desert-living hamadryas baboons:** Kummer (1995).

175 **Among the Bonerif:** Oosterwal (1961), pp. 99, 134.

175 **Marriage in the United States:** Browne (2002).

176 **"binds specific people":** Collier and Rosaldo (1981), p. 279.

176 **"an empty compliment":** Mill (1966 [1869]), p. 518. Millett (1970) reviews the Victorian debate between Mill and Ruskin (1902 [1865]).

Eight: The Cook's Journey

179 **Our long life spans suggest that our ancestors were good at escaping predators:** Safer species living longer: Austad and Fischer (1991). Reznick et al. (2004) show that the relationship is not necessarily straightforward.

180 **a faster rate of growth for the young:** The expected rate of growth in *Homo erectus* is a complex matter, and the fossil data are confusing (Aiello and Wells [2002], Moggi-Cecchi [2001]). Dean et al. (2001) showed the thickness of tooth enamel increased per day in early *Homo* at the same rate as in African apes, and concluded that *erectus* teeth grew at the same rate as in apes, although faster than *habilis.* They suggested that this meant that *erectus* had fast (apelike) rates of body growth. An infant *Homo erectus* from Indonesia likewise supports the idea of rapid growth. It was estimated from its cranial sutures to be only one year old when it died, yet it had already completed most of its brain growth. This indicated a rapid rate of growth similar to that of a chimpanzee, much faster than in *Homo sapiens* (Coqueugniot et al. [2004]). In contrast, Smith (1991) showed that data on the timing of the emergence of the third molar (whose appearance is considered to end the juvenile period) gave *habilis* a growth pattern like australopithecines, whereas *erectus* had a growth pattern like *Homo sapiens.* Clegg and

Aiello (1999) combined skeletal and dental analysis to suggest that *Homo erectus* (based on WT 15000) had a rate of growth within the range of *sapiens*. The debate continues (Anton 2003). Note that the set of life-history data that I predict here from considering *erectus* to have controlled fire and cooked are almost identical to those predicted by Hawkes et al. (1998) as resulting from grandmothers helping their daughters to mother. Fire and grandmothering could have worked side by side, and it is not clear which would have been a more important influence on growth, birth rates, and life span. Low weaning age in humans: Low (2000). Although the availability of weaning foods should have increased juvenile growth rates, slower growth is expected as a result of larger brains and longer lives, allowing energy to be diverted to the immune system and other defenses. Larger brains in longer-lived primates: Kaplan and Robson (2002). Investment in immune system correlated with longevity: Rolff (2002) and Nunn et al. (2008) show some evidence for this still little-understood relationship.

180 **The advantages of help given by grandmothers:** Hrdy (1999) and Hawkes et al. (1998) discuss the importance of cooperation in hunter-gatherer families.

180 **the thrifty-gene hypothesis:** Wells (2006) reviews the idea of the "thrifty-gene hypothesis," suggesting that humans are physiologically adapted to an erratic food supply. He implies that great apes are not subject to significant seasonal variation in food supply, which is clearly untrue (Pusey et al. [2005]). As Pond (1998) notes, humans lose relatively little body fat during seasonal food shortage compared to tropical animals of similar size.

182 **Every species other than humans can maintain adequate body heat without fire:** Darwin appears to have thought of fire as an adaptive response to cold. In his discussion of humans' ability to adapt to new conditions, he wrote, "When he migrates into a colder climate he uses clothes, builds sheds, and makes fires; and by the aid of fire cooks food otherwise indigestible" (Darwin [1871], chapter 6). Although the first

fire-users did not need fire, they could have benefited energetically from it (Pullen [2005]).

182 **Humans are exceptional runners:** Bramble and Lieberman (2004).

183 ***Homo erectus* could have lost their hair only if:** Wheeler (1992) explained human hair loss as a way to lose heat but did not discuss the use of fire to solve thermoregulation at night. Pagel and Bodmer (2003) note that fire would have solved the problem of maintaining warmth when inactive, but argued that the benefit of the loss of hair would have been reduced vulnerability to parasites, rather than allowing increased rates of heat loss by day.

183 **human babies are unique:** Kuzawa (1998) notes that although the exceptionally thick fat layer of human babies is commonly assumed to function as insulation to compensate for hairlessness, it serves additional functions, such as providing energy to fight infection or to tide babies over during periods of food shortage. Human infants gain fat shortly before birth and average 15 percent fat, compared to 1 percent to 2 percent in most mammals. Pond (1998) argues that although humans are often hypothesized to be relatively fat as adults, there is much evidence against the idea that fat insulates us as adults. Human fat concentrations are about the same in all climates and are not located in parts of the body that are effective for insulation.

184 **Raymond and Lorna Coppinger:** Coppinger and Coppinger (2000).

185 **In animals, more tolerant individuals cooperate:** Chimpanzees more tolerant: Melis et al. (2006a, 2006b). Bonobos more tolerant: Hare et al. (2007). Foxes more tolerant: Hare et al. (2005).

186 **"expectations of men's entitlement":** DeVault (1997), p. 180.

186 **a primal urge to quench the flames:** cited in Goudsblom (1992), p. 19.

187 ***Paranthropus* relied mainly on a diet:** Sponheimer et al. (2006).

188 **lions and saber-tooths:** Werdelin and Lewis (2005) review the predators living in Africa during early human and pre-human evolution.

188 **the short sticks chimpanzees use:** Pruetz and Bertolani (2007).

188 **chimpanzees now sometimes scare pigs or humans with well-aimed missiles:** Goodall (1986).

188 **If they threw rocks:** Toth and Schick (2006) review the use of rocks in the earliest stone age, from 2.6 million years ago.

189 **Knowing that habilines were able to cut steaks:** Overview of habiline feeding strategies: Perlès (1999), Dominguez-Rodrigo (2002), Ungar (2006). Plummer (2004) discusses habilines and *Homo erectus* in relation to tools and diet.

190 **children as young as two years old make their own fires:** Goudsblom (1992, p. 197) cites anecdotes of two-and three-year-olds making their own fires from their mothers' fires among both Tiwi and Kung !San.

190 **chimpanzees and bonobos can tend fires:** Brewer (1978, pp. 174–176) described the behavior of chimpanzees that were being rehabilitated into the wild in Senegal. They managed camp fires in a rudimentary way for cooking and warmth. Raffaele (2006) mentions fire-making by Kanzi, the bonobo studied by Savage-Rumbaugh (Savage-Rumbaugh and Lewin [1994]). Brink (1957) describes chimpanzees in Johannesburg Zoo chain-smoking cigarettes by continuously lighting them.

190 **Sparks produced by accident from pounded rocks:** Darwin (1871), p. 808. The Oldowan stone-tool culture that habilines must have used includes numerous fist-size hammerstones that could well have served the purpose of tenderizing meat (Mora and de la Torre [2005]).

191 **Yakuts of Siberia:** Frazer (1930), p. 226.

191 **tinder fungus:** Tinder fungus: survival manuals recommend catching fire on species of the genus *Fomes*, because after sparks land on the dry bracket fungus, they spread slowly in a widening ring, staying alight a long time (e.g., *www.wildwood survival.com/survival/fire/twostones*). The preferred species,

Fomes fomentarius, is common in East Africa. It holds fire so well that Osage Indians of North America kept fire for several days by taking tinder fungus from inside a hollow tree, lighting it, enclosing it in earth, placing it between the two valves of a hollow mussel shell, and wrapping and binding it with cord (Hough [1926], p. 3).

191 **Anthropologists caution:** Oakley (1955), Collin et al. (1991). Rowlett (1999) reported that chert artifacts good for starting fires were unusually numerous at Koobi Fora.

191 **standard components of fire-making kits:** Hough (1926), Frazer (1930).

192 **a eucalyptus tree can smolder:** Clark and Harris (1985).

192 **burning nonstop near Antalya:** The flames can be seen near Antalya in the shadow of Mount Olympus. Methane and other gases emerge from narrow slits in the rock many meters long, creating a cluster of "eternal" flames in the bare hillside. Homer described it as the place where the monster Chimera lit the earth with its dying breath. The flames appear to have declined in height over the past two thousand to three thousand years but show no sign of being extinguished.

193 **Among Australian aborigines:** Tindale (1974).

193 **When people stop, they start a small fire:** Turnbull (1962), p. 58, illustrates the pattern with Mbuti Pygmies of central Africa: "The first thing they do when they stop on the trail for a rest is to unwrap the ember and, putting some dry twigs around it, blow softly once or twice and transform it into a blazing fire." Basedow (1925), p. 110, describes a similar pattern for the Aranda of Australia: "Perhaps the most important article a native possesses is the fire-stick. No matter where he might be, on the march or in camp, it is his constant companion. Important as it is, the fire-stick is only a short length of dry branch or bark, smouldering at one end. It is carried in the hand with a waving motion, from one side to another. When walking in the dark, this motion is brisker in order to keep alive sufficient flame for lighting the way. A body of natives walking in this way at night, in the customary Indian file, is indeed an imposing sight. Di-

rectly a halt is made, a fire is lit, to cook the meals at day and to supply warmth during sleep at night. When camp is left, a fresh stick is taken from the fire and carried on to the next stopping place." Such behavior has been widely described in hunter-gatherers.

Epilogue: The Well-Informed Cook

195 **"overweight enough to begin experiencing health problems":** Critser (2003).

195 **as John Kenneth Galbraith first noted:** Galbraith (1958).

196 **The detail of biochemical knowledge:** Johnson (1994, 2001), Smith and Morton (2001).

201 **tweak the original Atwater system:** Southgate and Durnin (1970) extended Atwater's general factors; Southgate (1981) presented further modifications.

202 **our metabolic rate rises, the maximum increase averaging 25 percent:** Data on the costs of digestion and factors affecting it: Secor (2009).

202 **people eating a high-fat diet:** Sims and Danforth (1987).

203 **the costs of digestion are higher for tougher or harder foods than softer foods:** Secor (2009).

203 **for foods with larger than smaller particles:** Heaton et al. (1988).

203 **When A. L. Merrill and B. K. Watt introduced the Atwater specific-factor system:** Merrill and Watt (1955).

204 **many nutritionists have called for a major revision of the Atwater convention:** Livesey (2001) cites twenty-two expert reviews, reports, and regulatory documents calling for a change to the system for characterizing energy value in food labels. Collectively, these reports favor the view that the heat increment produced during digestion should be taken into account.

206 **Japanese women:** Murakami et al. (2007). See et al. (2007) showed that larger waists are associated with increased moratality.

206 **as food-writer Michael Pollan has argued:** Pollan (2008).

206 **We once thought of our species as infinitely adaptable:** Archaeologist Robert Kelly expresses a popular view: "There is no original human society, no basal human adaptation: studying modern hunter-gatherers in order to subtract the effects of contact with the world system (were that possible) and to uncover universal behaviors with the goal of reconstructing the original hunter-gatherer lifeway is simply not possible—because that lifeway never existed" (Kelly [1995], p. 337). Archaeologist Rick Potts illustrates the same idea. "It is patently incorrect," he says, "to characterize the human ancestral environment as a set of specific repetitive elements, statistical regularities, or uniform problems which the cognitive mechanisms unique to humans are designed to solve" (Potts [1998], pp. 129–130). Adaptation to the hearth suggests these kinds of views need to be modified.

BIBLIOGRAPHY

Agetsuma, N., and N. Nakagawa. 1998. "Effects of Habitat Differences on Feeding Behaviors of Japanese Monkeys: Comparison Between Yakushima and Kinkazan. *Primates* 39:275–289.

Aiello, L., and J. C. K. Wells. 2002. "Energetics and the Evolution of the Genus *Homo*." *Annual Review of Anthropology* 31:323–338.

Aiello, L., and P. Wheeler. 1995. "The Expensive-Tissue Hypothesis: The Brain and the Digestive System in Human and Primate Evolution." *Current Anthropology* 36:199–221.

Albert, R. M., O. Bar-Yosef, L. Meignen, and S. Weiner. 2003. "Quantitative Phytolith Study of Hearths from the Natufian and Middle Palaeolithic Levels of Hayonim Cave (Galilee, Israel)." *Journal of Archaeological Science* 30:461–480.

Alberts, S. C., H. E. Watts, and J. Altmann. 2003. "Queuing and Queue-Jumping: Long Term Patterns of Reproductive Skew Among Male Savannah Baboons." *Animal Behavior* 65:821–840.

Alexander, R. D. 1987. *The Biology of Moral Systems.* Hawthorne, NY: Aldine de Gruyter.

———. 1990. "How Did Humans Evolve? Reflections on the Uniquely Unique Species." *Museum of Zoology, The University of Michigan, Special Publication* 1:1–40.

Alperson-Afil, N. 2008. "Continual Fire-Making by Hominins at Gesher Benot Ya'aqov, Israel." *Quaternary Science Reviews* 27:1733–1739.

Antón, S. C. 2003. "Natural History of *Homo Erectus*." *Yearbook of Physical Anthropology* 46:126–170.

Antón, S. C., and C. C. I. Swisher. 2004. "Early Dispersals of *Homo* from Africa." *Annual Review of Anthropology* 33:271–296.

Arlin, S., F. Dini, and D. Wolfe. 1996. *Nature's First Law: the Raw-Food Diet.* San Diego: Maul Brothers.

Armbrust, L. J., J. J. Hoskinson, M. Lora-Michiels, and G. A. Milliken. 2003. "Gastric Emptying in Cats Using Foods Varying in Fiber Content and Kibble Shapes." *Veterinary Radiology and Ultrasound* 44:339–343.

Arnqvist, G., T. M. Jones, and M. A. Elgar. 2006. "Sex-Role Reversed Nuptial Feeding Reduces Male Kleptoparasitism of Females in Zeus Bugs (Heteroptera; Veliidae)." *Biology Letters* 2:491–493.

Atkins, P., and I. Bowler. 2001. *Food in Society: Economy, Culture, Geography.* London: Arnold.

Austad, S. N., and K. E. Fischer. 1991. "Mammalian Aging, Metabolism, and Ecology—Evidence from the Bats and Marsupials." *Journal of Gerontology* 46:B47–B53.

Baksh, M. 1990. *Time Allocation Among the Machiguenga of Camaná.* New Haven, CT: Human Relations Area Files Inc.

Barham, P. 2000. *The Science of Cooking.* Berlin: Springer.

Barr, S. I. 1999. "Vegetarianism and Menstrual Cycle Disturbances: Is There an Association?" *American Journal of Clinical Nutrition* 70:549S–554S.

Barton, R. A. 1992. "Allometry of Food Intake in Free-Ranging Anthropoid Primates." *Folia primatologica* 58:56–59.

Barton, R. N. E., A. P. Currant, Y. Fernandez-Jalvo, J. C. Finlayson, P. Goldberg, R. Macphail, P. B. Pettitt, and C. B. Stringer. 1999. "Gibralter Neanderthals and Results of Recent Excavations in Gorham's, Vanguard and Ibex Caves." *Antiquity* 73:13–23.

Basedow, H. 1925. *The Australian Aboriginal.* Adelaide, Australia: F. W. Preece.

Beaumont, W. 1996 (first published 1833). *Experiments and Observations on the Gastric Juice and the Physiology of Digestion.* Mineola, NY: Dover.

Becker, G. S. 1985. "Human Capital, Effort, and the Sexual Division of Labor." *Journal of Labor Economics* 3:S33–S58.

Beeton, I. 1909. *Mrs. Beeton's Book of Household Management.* London: Ward, Lock.

Bermudez de Castro, J. M., and M. E. Nicolas. 1995. "Posterior Dental Size Reduction in Hominids: The Atapuerca Evidence." *American Journal of Physical Anthropology* 96:335–356.

Berndt, R. M., and C. H. Berndt. 1988. *The World of the First Australians.* Canberra, Australia: Aboriginal Studies Press.

Bird, R. 1999. "Cooperation and Conflict: The Behavioral Ecology of the Sexual Division of Labor." *Evolutionary Anthropology* 8:65–75.

Boag, P. T., and P. R. Grant. 1981. "Intense Natural Selection in a Population of Darwin's Finches (Geospizinae) in the Galápagos." *Science* 214:82–85.

Boback, S. M. 2006. "A Morphometric Comparison of Island and Mainland Boas (*Boa constrictor*) in Belize." *Copeia*:261–267.

Boback, S. M., C. L. Cox, B. D. Ott, R. Carmody, R. W. Wrangham, and S. M. Secor. 2007. "Cooking Reduces the Cost of Meat Digestion." *Comparative Biochemistry and Physiology* 148:651–656.

Boehm, C. 1999. *Hierarchy in the Forest: The Evolution of Egalitarian Behavior.* Cambridge, MA: Harvard University Press.

Boyd, R., and J. B. Silk. 2002. *How Humans Evolved.* New York: W. W. Norton.

Brace, C. L. 1995. *The Stages of Human Evolution*, 5th ed. Englewood Cliffs, NJ: Prentice-Hall.

Bramble, D. M., and D. E. Lieberman. 2004. "Endurance Running and the Evolution of Homo." *Nature* 432:345–352.

Brand-Miller, J. 2006. *The New Glucose Revolution. New York:* Da Capo Press.

Breeks, J. W. 1873. *An Account of the Primitive Tribes and Monuments of the Nilagiris.* London: W. H. Allen.

Brewer, S. 1978. *The Forest Dwellers.* London: Collins.

Bricker, H. M. 1995. *Le Paleolithique Superieur de l'Abri Pataud (Dordogne): Les Fouilles de H. L. Movius, Jr.* Paris: Documents d'Archéologie Française, Maison des Sciences de l'Homme.

Brillat-Savarin, J. A. 1971. *The Physiology of Taste: Or Meditations on Transcendental Gastronomy (1825).* New York: Alfred A. Knopf.

Brink, A. 1957. "The Spontaneous Fire-Controlling Reactions of Two Chimpanzee Smoking Addicts." *South African Journal of Science* 53:241–247.

Brown, M. A., L. H. Storlien, I. L. Brown, and J. A. Higgins. 2003. "Cooking Attenuates the Ability of High-Amylose Meals to Reduce Plasma Insulin Concentrations in Rats." *British Journal of Nutrition* 90:823–827.

Browne, K. 2002. *Biology at Work: Rethinking Sexual Equality.* New Brunswick, NJ: Rutgers University Press.

Bunn, H. T., and C. B. Stanford. 2001. "Research Trajectories and Hominid Meat-Eating." In *Meat-Eating and Human Evolution,* C. B. Stanford and H. T. Bunn, eds., 350–359. New York: Oxford University Press.

Burch, E. 1998. *The Inupiaq Eskimo Nations of Northwest Alaska.* Fairbanks: University of Alaska Press.

Byrne, R. W., and L. A. Bates. 2007. "Sociality, Evolution and Cognition." *Current Biology* 17: R714–R723.

Campling, R. C. 1991. "Processing Grains for Cattle—a Review." *Livestock Production Science* 28:223–234.

Carmody, R., and R. W. Wrangham. At press. "The Energetic Significance of Cooking." *Journal of Human Evolution.*

Carpenter, J. E., and S. Bloem. 2002. "Interaction Between Insect Strain and Artificial Diet in Diamondback Moth Development and Reproduction." *Entomologia Experimentalis et Applicata* 102:283–294.

Cartmill, M. 1993. *A View to a Death in the Morning: Hunting and Nature through History.* Cambridge, MA: Harvard University Press.

Charnov, E. L. 1993. *Life-History Invariants: Some Explorations of Symmetry in Evolutionary Ecology.* Oxford, UK: Oxford University Press.

Chivers, D. J., and C. M. Hladik. 1980. "Morphology of the Gastrointestinal Tract in Primates: Comparison with Other Mammals in Relation to Diet." *Journal of Morphology* 166:337–386.

———. 1984. "Diet and Gut Morphology in Primates." In *Food Acquisition and Processing in Primates,* D. J. Chivers, B. A. Wood, and A. Bilsborough, eds., 213–230. New York: Plenum Press.

Christian, M. G., and Christian, E. 1904. *Uncooked Foods and How to Use Them: A Treatise on How to Get the Highest Form of Animal Energy from Food.* New York: The Health-Culture Company.

Clark, J. D., and J. W. K. Harris. 1985. "Fire and Its Role in Early Hominid Lifeways." *African Archaeological Review* 3:3–27.

Clegg, M., and L. C. Aiello. 1999. "A Comparison of the Nariokotome *Homo erectus* with Juveniles from a Modern Human Population." *American Journal of Physical Anthropology* 110:81–94.

Clutton-Brock, T. H., and P. H. Harvey. 1977. "Species Differences in Feeding and Ranging Behaviour in Primates." In *Primate Ecology,* T. H. Clutton-Brock, ed., 557–580. London: Academic Press.

Cnotka, J., O. Güntürkün, G. Rehkämper, R. D. Gray, and G. R. Hunt. 2008. "Extraordinary Large Brains in Tool-Using New Caledonian Crows *(Corvus moneduloides)*." *Neuroscience Letters* 433:241–245.

Cohn, E. W. 1936. "In Vitro and In Vivo Experiments on the Digestibility of Heat-Treated Egg White." PhD diss., University of Chicago.

Collard, M., and B. A. Wood. 1999. "Grades Among the African Early Hominids." In *African Biogeography, Climate Change, and Early Hominid Evolution,* T. Bromage and F. Schrenk, eds., 316–327. New York: Oxford University Press.

Collier, J. F., and M. Z. Rosaldo. 1981. "Politics and Gender in Simple Societies." In *Sexual Meanings: The Cultural Construction of Gender and Sexuality,* S. B. Ortner and H. Whitehead, eds., 275–329. Cambridge, UK: Cambridge University Press.

Collin, F., D. Mattart, L. Pirnay, and J. Speckens. 1991. "L'obtention du feu par percussion: approche experimentale et traceologique." *Bulletin des Chercheurs de la Wallonie* 31:19–49.

Collings, P., C. Williams, and I. MacDonald. 1981. "Effects of Cooking on Serum Glucose and Insulin Responses to Starch." *British Medical Journal* 282:1032.

Combes, S., J. Lepetit, B. Darche, and F. Lebas. 2003. "Effect of Cooking Temperature and Cooking Time on Warner-Bratzler Tenderness Measurement and Collagen Content in Rabbit Meat." *Meat Science* 66:91–96.

Conklin-Brittain, N., R. W. Wrangham, and C. C. Smith. 2002. "A Two-Stage Model of Increased Dietary Quality in Early Hominid Evolution: The Role of Fiber." In *Human Diet: Its Origin and Evolution,* P. Ungar and M. Teaford, eds., 61–76. Westport, CT: Bergin & Garvey.

Connor, R. C. 2007. "Dolphin Social Intelligence: Complex Alliance Relationships in Bottlenose Dolphins and a Consideration of Selective Environments for Extreme Brain Size Evolution in Mammals." *Philosophical Transactions of the Royal Society of London Series B* 362:587–602.

Coon, C. S. 1962. *The History of Man: From the First Human to Primitive Culture and Beyond.*, 2nd ed. London: Jonathan Cape.

Coppinger, R., and L. Coppinger. 2000. *Dogs: A Startling New Understanding of Canine Origin, Behavior, and Evolution.* New York: Scribner.

Coqueugniot, H., J.-J. Hublin, F. Veillon, F. Houet, and T. Jacob. 2004. "Early Brain Growth in *Homo erectus* and Implications for Cognitive Ability." *Nature* 431:299–302.

Critser, G. 2003. *Fat Land: How Americans Became the Fattest People in the World.* Boston, MA: Houghton Mifflin.

Darwin, C. 1871 (2006). *The Descent of Man, and Selection in Relation to Sex.* In *From So Simple a Beginning: The Four Great Books of Charles Darwin,* E. O. Wilson, ed. New York: W. W. Norton, pp. 767–1254.

———. 1888. *A Naturalist's Voyage. Journal of Researches into the Natural History and Geology of the Countries Visited During the Voyage of H.M.S. "Beagle" Round the World Under the Command of Capt. Fitzroy, R.N.*, 3rd ed. London: John Murray.

Davies, K. J. A., S. W. Lin, and R. E. Pacifici. 1987. "Protein Damage and Degradation by Oxygen Radicals. IV. Degradation of Denatured Protein." *Journal of Biological Chemistry* 262:9914–9920.

Dawson, J. 1881. *Australian Aborigines: The Languages and Customs of Several Tribes of Aborigines in the Western District of Victoria, Australia.* Melbourne, Australia: George Robertson.

de Araujo, I. E., and E. T. Rolls. 2004. "Representations in the Human Brain of Food Texture and Oral Fat." *Journal of Neuroscience* 24:3086–3093.

de Huidobro, F. R., E. Miguel, B. Blazquez, and E. Onega. 2005. "A Comparison Between Two Methods (Warner-Bratzler and Texture Profile Analysis) for Testing Either Raw Meat or Cooked Meat." *Meat Science* 69:527–536.

Dean, C., M. G. Leave, D. Reid, F. Schrenk, G. T. Schwartz, C. Stringer, and A. Walker. 2001. "Growth Processes in Teeth Distinguish Modern Humans from *Homo Erectus* and Earlier Hominins." *Nature* 414:628–631.

Deaner, R. O., K. Isler, J. Burkart, and C. van Schaik. 2007. "Overall Brain Size, and Not Encephalization Quotient, Best Predicts Cognitive Ability Across Non-Human Primates." *Brain, Behavior and Evolution* 70:115–124.

DeGusta, D., H. W. Gilbert, and S. P. Turner. 1999. "Hypoglossal Canal Size and Hominid Speech." *Proceedings of the National Academy of Sciences* 96:1800–1804.

DeVault, M. 1997. "Conflict and Deference." In *Food and Culture: A Reader,* C. Counihan and P. van Esterik, eds., 180–199. New York: Routledge.

Devivo, R., and A. Spors. 2003. *Genefit Nutrition.* Berkeley, CA: Celestial Arts.

Dominguez-Rodrigo, M. 2002. "Hunting and Scavenging by Early Humans: The State of the Debate." *Journal of World Prehistory* 16:1–54.

Donaldson, M. S. 2001. "Food and Nutrient Intake of Hallelujah Vegetarians." *Nutrition and Food Science* 31:293–303.

Doran, D. M., and A. McNeilage. 1998. "Gorilla Ecology and Behavior." *Evolutionary Anthropology* 6:120–131.

Driver, H. E. 1961. *Indians of North America.* Chicago: University of Chicago Press.

Dunbar, R. I. M. 1998. "The Social Brain Hypothesis." *Evolutionary Anthropology* 6:178–190.

Durkheim, E. 1933. *On the Division of Labor in Society.* George Simpson, trans. New York: Macmillan.

Dzudie, T., R. Ndjouenkeu, and A. Okubanjo. 2000. "Effect of Cooking Methods and Rigor State on the Composition, Tenderness and Eating Quality of Cured Goat Loins." *Journal of Food Engineering* 44:149–153.

Eastwood, M. 2003. *Principles of Human Nutrition*, 2nd ed. Oxford, UK: Blackwell.

Ellison, P. 2001. *On Fertile Ground.* Cambridge, MA: Harvard University Press.

Emmons, G. T. 1991. *The Tlingit Indians.* Seattle: University of Washington Press.

Engelen, L., R. A. de Wijk, A. van der Bilt, J. F. Prinz, A. M. Janssen, and F. Bosman. 2005a. "Relating Particles and Texture Perception." *Physiology and Behavior* 86:111–117.

Engelen, L., A. Fontijn-Tekamp, and A. van der Bilt. 2005b. "The Influence of Product and Oral Characteristics on Swallowing." *Archives of Oral Biology* 50:739–746.

Englyst, H. N., and J. H. Cummings. 1985. "Digestion of the Polysaccharides of Some Cereal Foods in the Human Small Intestine." *American Journal of Clinical Nutrition* 42:778–787.

———. 1986. "Digestion of the Carbohydrates of Banana (*Musa paradisiaca sapientum*) in the Human Small Intestine." *American Journal of Clinical Nutrition* 444:42–50.

———. 1987. "Digestion of Polysaccharides of Potato in the Small Intestine of Man." *American Journal of Clinical Nutrition* 45:423–431.

Evenepoel, P., D. Claus, B. Geypens, M. Hiele, K. Geboes, P. Rutgeerts, and Y. Ghoos. 1999. "Amount and Fate of Egg Protein Escaping Assimilation in the Small Intestine of Humans." *American Journal of Physiology (Endocrinol. Metabol.)* 277:G935–G943.

Evenepoel, P., B. Geypens, A. Luypaerts, M. Hiele, and P. Rutgeerts. 1998. "Digestibility of Cooked and Raw Egg Protein in Humans as Assessed by Stable Isotope Techniques." *Journal of Nutrition* 128:1716–1722.

Felger, R., and M. B. Moser. 1985. *People of the Desert and Sea: Ethnobotany of the Seri Indians.* Tucson: University of Arizona Press.

Fernández-Armesto, F. 2001. *Food: A History*. London: Macmillan.

Fish, J. L., and C. A. Lockwood. 2003. "Dietary Constraints on Encephalization in Primates." *American Journal of Physical Anthropology* 120:171–181.

Fisher, J. R., and D. J. Bruck. 2004. "A Technique for Continuous Mass Rearing of the Black Vine Weevil, *Otiorhyncus Sulcatus*." *Entomologia Experimentalis et Applicata* 113:71–75.

Foley, R. 2002. "Adaptive Radiations and Dispersals in Hominin Evolutionary Ecology." *Evolutionary Anthropology* 11:32–37.

Fontana, B. L. 2000. *Trails to Tiburon: The 1894 and 1895 Field Diaries of W. J. McGee.* Tucson: University of Arizona Press.

Fontana, L., J. L. Shew, J. O. Holloszy, and D. T. Villareal. 2005. "Low Bone Mass in Subjects on a Long-Term Raw Vegetarian Diet." *Archives of Internal Medicine* 165:684–689.

Food Standards Agency. 2002. *McCance and Widdowson's The Composition of Foods: Sixth Summary Edition*. Cambridge, UK: Royal Society of Chemistry.

Frazer, J. G. 1930 (reprinted 1974). *Myths of the Origins of Fire*. New York: Hacker Art Books.

Fry, T. C., H. M. Shelton, and D. Klein. 2003. *Self Healing Power! How to Tap into the Great Power Within You*. Sebastopol, CA: Living Nutrition.

Fuentes, A. 2000. "Hylobatid Communities: Changing Views on Pair Bonding and Social Organization in Hominoids." *Yearbook of Physical Anthropology* 43:33–60.

Fullerton-Smith, J. 2007. *The Truth About Food: What You Eat Can Change Your Life*. London: Bloomsbury.

Funston, P. J., M. G. L. Mills, H. C. Biggs, and P. R. K. Richardson. 1998. "Hunting by Male Lions: Ecological Implications and Socioecological Influences." *Animal Behavior* 56:1333–1345.

Galbraith, J. K. 1958. *The Affluent Society*. Boston: Houghton Mifflin.

Gaman, P. M., and K. B. Sherrington. 1996. *The Science of Food: An Introduction to Food Science, Nutrition and Microbiology*. Oxford, UK: Pergamon Press.

Gilby, I. C. 2006. "Meat Sharing Among the Gombe Chimpanzees: Harassment and Reciprocal Exchange." *Animal Behaviour* 71:953–963.

Gilby, I. C., L. E. Eberly, L. Pintea, A. E. Pusey. 2006. "Ecological and Social Influences on the Hunting Behaviour of Wild Chimpanzees, *Pan troglodytes schweinfurthii*." *Animal Behaviour* 72:169–180.

Gilby, I. C., and R. Wrangham. 2007. "Risk-Prone Hunting by Chimpanzees (*Pan troglodytes schweinfurthii*) Increases During Periods of High Diet Quality." *Behavioral Ecology and Sociobiology* 61:1771–1779.

Gilman, C. P. 1966 (1898). *Women and Economics: A Study of the Economic Relation Between Men and Women as a Factor in Social Evolution*. New York: Harper.

Gladwin, T., and S. B. Sarason. 1953. "Truk: Man in Paradise." *Viking Fund Publications in Anthropology* 29:1–655.

Goodall, J. 1986. *The Chimpanzees of Gombe: Patterns of Behavior*. Cambridge, MA: Harvard University Press.

———. 1982. *Cooking, Cuisine and Class: A Study in Comparative Sociology*. Cambridge, UK: Cambridge University Press.

Goren-Inbar, N., N. Alperson, M. E. Kislev, O. Simchoni, Y. Melamed, A. Ben-Nun, and E. Werker. 2004. "Evidence of Hominin Control of Fire at Gesher Benot Ya'aqov, Israel." *Science* 304, 725–727.

Gott, B. 2002. "Fire-Making in Tasmania: Absence of Evidence is Not Evidence of Absence." *Current Anthropology* 43:650–656.

Goudsblom, J. 1992. *Fire and Civilization.* New York: Penguin.

Gould, S. J. 2002. *The Structure of Evolutionary Theory.* Cambridge, MA: Harvard University Press.

Gowlett, J. A. J. 2006. "The Early Settlement of Northern Europe: Fire History in the Context of Climate Change and the Social Brain." *C. R. Palevol* 5:299–310.

Gowlett, J. A. J., J. Hallos, S. Hounsell, V. Brant, and N. C. Debenham. 2005. "Beeches Pit—Archaeology, Assemblage Dynamics and Early Fire History of a Middle Pleistocene Site in East Anglia, UK." *Journal of Eurasian Archaeology* 3:3–40.

Grant, P. R., and B. R. Grant. 2002. "Unpredictable Evolution in a 30-year Study of Darwin's Finches." *Science* 296:707–711.

Gregor, T. 1985. *Anxious Pleasures: The Sexual Lives of an Amazonian People.* Chicago: University of Chicago Press.

Grinker, R. R. 1994. *Houses in the Rain Forest: Ethnicity and Inequality Among Farmers and Foragers in Central Africa.* Berkeley: University of California Press.

Gusinde, M. 1961. *The Yamana: The Life and Thought of the Water Nomads of Cape Horn.* Frieda Schutze, trans. New Haven, CT: Human Relations Area Files.

Haeusler, M., and H. M. McHenry. 2004. "Body Proportions of *Homo Habilis* Reviewed." *Journal of Human Evolution* 46:433–465.

Hagen, A. 1998. *A Handbook of Anglo-Saxon Food: Processing and Consumption.* Hockwold-cum-Wilton, Norfolk, UK: Anglo-Saxon Books.

Hames, R. 1993. *Ye'kwana Time Allocation.* New Haven, CT: Human Relations Area Files Inc.

Hamilton, A. 1987. "Dual Social System: Technology, Labour and Women's Secret Rites in the Eastern Western Desert of Australia." In *Traditional Aboriginal Society: A Reader,* W. H. Edwards, ed., 34–52. Melbourne, Australia: Macmillan.

Handy, E. S. C. 1923. "The Native Culture in the Marquesas." *Bernice P. Bishop Museum Bulletin* 9:1–358.

Hare, B., A. P. Melis, V. Woods, S. Hastings, and R. Wrangham. 2007. "Tolerance Allows Bonobos to Outperform Chimpanzees on a Cooperative Task." *Current Biology* 17:619–623.

Hare, B., I. Plyusnina, N. Ignacio, O. Schepina, A. Stepika, R. Wrangham, and L. Trut. 2005. "Social Cognitive Evolution in Captive Foxes Is a Correlated By-Product of Experimental Domestication." *Current Biology* 15:1–20.

Harris, P. V., and W. R. Shorthose. 1988. "Meat Texture." In *Developments in Meat Science,* R. A. Lawrie, ed., 245–296. London: Elsevier.

Hart, C. W. M., and A. R. Pilling. 1960. *The Tiwi of North Australia.* New York: Holt, Rinehart and Winston.

Hawk, P. B. 1919. *What We Eat and What Happens to It: The Results of the First Direct Method Ever Devised to Follow the Actual Digestion of Food in the Human Stomach.* New York: Harper.

Hawkes, K., J. O'Connell, and N. Blurton-Jones. 1997. "Hadza Women's Time Allocation, Offspring Provisioning, and the Evolution of Long Menopausal Lifespans." *Current Anthropology* 38:551–577.

———. 2001a. "Hadza Meat Sharing." *Evolution and Human Behavior*, 22, 113–142.

———. 2001b. "Hunting and Nuclear Families: Some Lessons from the Hadza About Men's Work." *Current Anthropology* 42:681–709.

Hawkes, K., J. F. O'Connell, N. G. Blurton-Jones, H. Alvarez, and E. L. Charnov. 1998. "Grandmothering, Menopause, and the Evolution of Human Life Histories." *Proceedings of the National Academy of Sciences, USA* 95:1336–1339.

Headland, T. N., and L. A. Reid. 1989. "Hunter-Gatherers and Their Neighbors from Prehistory to the Present." *Current Anthropology* 30:27–43.

Heaton, K. W., S. N. Marcus, P. M. Emmett, and C. H. Bolton. 1988. "Particle Size of Wheat, Maize, and Oat Test Meals: Effects on Plasma Glucose and Insulin Responses and on the Rate of Starch Digestion In Vitro." *American Journal of Clinical Nutrition* 47:675–682.

Hernandez-Aguilar, R. A., J. Moore, and T. R. Pickering. 2007. "Savanna Chimpanzees Use Tools to Harvest the Underground Storage Organs of Plants." *Proceedings of the National Academy of Sciences* 104:19210–19213.

Heyerdahl, T. 1996. *The Kon-Tiki Expedition: By Raft Across the South Seas.* London: Flamingo.

Hiiemae, K. M., and J. B. Palmer. 1999. "Food Transport and Bolus Formation During Complete Feeding Sequences on Foods of Different Initial Consistency." *Dysphagia* 14:31–42.

Hladik, C. M., D. J. Chivers, and P. Pasquet. 1999. "On Diet and Gut Size in Non-Human Primates and Humans: Is There a Relationship to Brain Size?" *Current Anthropology* 40:695–697.

Hobbs, S. H. 2005. "Attitudes, Practices, and Beliefs of Individuals Consuming a Raw Foods Diet." *Explore* 1:272–277.

Hofferth, S. L., and J. F. Sandberg. 2001. "How American Children Spend Their Time." *Journal of Marriage and the Family* 63:295–308.

Hohmann, G., and B. Fruth. 2000. "Use and Function of Genital Contacts Among Female Bonobos." *Animal Behavior* 60:107–120.

Holekamp, K. E., S. T. Sakai, and B. L. Lundrigan. 2007. "Social Intelligence in the Spotted Hyena (*Crocuta crocuta*)." *Philosophical Transactions of the Royal Society of London, Series B* 362:523–538.

Holmberg, A. R. 1969. *Nomads of the Longbow: The Siriono of Eastern Bolivia.* Garden City, NY: Natural History Press.

Hough, W. 1926. *Fire as an Agent in Human Culture.* Washington, DC: U.S. Government Printing Office.

Howell, E. 1994. *Food Enzymes for Health and Longevity.* Twin Lakes, WI: Lotus Press.

Hrdy, S. B. 1999. *Mother Nature: A History of Mothers, Infants, and Natural Selection.* New York: Pantheon.

Hunt, K. D. 1991. "Positional Behavior in the Hominoidea." *International Journal of Primatology* 12:95–118.

Hunt, P. 1961. *Eating and Drinking: An Anthology for Epicures.* London: Ebury Press.

Hurtado, J. L., P. Montero, J. Borderias, and M. T. Solas. 2001. "Morphological and Physical Changes During Heating of Pressurized Common Octopus Muscle up to Cooking Temperature." *Food Science and Technology International* 7:329–338.

Isaacs, J. 1987. *Bush Food: Aboriginal Food and Herbal Medicine.* Sydney, Australia: New Holland.

Isler, K., and C. P. van Schaik. 2006. "Costs of Encephalization: The Energy Trade-Off Hypothesis Tested on Birds." *Journal of Human Evolution* 51:228–243.

James, S. R. 1989. "Hominid Use of Fire in the Lower and Middle Pleistocene: A Review of the Evidence." *Current Anthropology* 30:1–26.

Jenike, M. 2001. "Nutritional Ecology: Diet, Physical Activity and Body Size." In *Hunter-Gatherers: An Interdisciplinary Perspective,* C. Panter-Brick, R. H. Layton, and P. Rowley-Conwy, eds., 205–238. Cambridge, UK: Cambridge University Press.

Jenkins, D. J. A. 1988. "Nutrition and Diet in Management of Diseases of the Gastrointestinal Tract. (C) Small Intestine: (6) Factors Influencing Absorption of Natural Diets." In *Modern Nutrition in Health and Disease,* M. E. Shils, and V. R. Young, eds., 1151–1166. Philadelphia: Lea and Febiger.

Jenness, D. 1922. *Report of the Canadian Arctic Expedition 1913–18. Volume XII: The Life of the Copper Eskimos.* Ottawa: F. A. Acland.

Johnson, A. 1975. "Time Allocation in a Machiguenga Community." *Ethnology* 14:301–310.

———. 2003. *Families of the Forest: The Matsigenka Indians of the Peruvian Amazon.* Berkeley, CA: University of California Press.

Johnson, A., and O. R. Johnson. 1988. *Time Allocation Among the Machiguenga of Shimaa.* New Haven, CT: Human Relations Area Files Inc.

Johnson, L. R. 1994. *Physiology of the Gastrointestinal Tract*, 3rd ed. New York: Raven Press.

———. 2001. *Gastrointestinal Physiology*, 6th ed. St. Louis, MO: Mosby.

Jolly, C., and R. White. 1995. *Physical Anthropology and Archaeology.* New York: McGraw-Hill.

Jones, M. 2007. *Feast: Why Humans Share Food.* New York: Oxford University Press.

Kaberry, P. M. 1939. *Aboriginal Woman: Sacred and Profane.* London: Routledge.

Kadohisa, M., E. T. Rolls, and J. V. Verhagen. 2004. "Orbitofrontal Cortex: Neuronal Representation of Oral Temperature and Capsaicin in Addition to Taste and Texture." *Neuroscience* 127:207–221.

———. 2005a. "Neuronal Representations of Stimuli in the Mouth: The Primate Insular Taste Cortex, Orbitofrontal Cortex and Amygdala." *Chemical Senses* 30:401–419.

Kadohisa, M., J. V. Verhagen, and E. T. Rolls. 2005b. "The Primate Amygdala: Neuronal Representations of the Viscosity, Fat Texture, Temperature, Grittiness and Taste of Foods." *Neuroscience* 132:33–48.

Kaplan, H., K. Hill, J. Lancaster, and A. M. Hurtado. 2000. "A Theory of Human Life History Evolution: Diet, Intelligence and Longevity." *Evolutionary Anthropology* 9:156–185.

Kaplan, H. S., and A. J. Robson. 2002. "The Emergence of Humans: The Coevolution of Intelligence and Longevity with Intergenerational Transfers." *Proceedings of the National Academy of Sciences* 99:10221–10226.

Karlsson, M. E., and A.-C. Eliasson. 2003. "Effects of Time/Temperature Treatments on Potato (*Solanum Tuberosum*) Starch: A Comparison of Isolated Starch and Starch *In Situ*." *Journal of the Science of Food and Agriculture* 83:1587–1592.

Kaufman, J. A. 2006. "On the Expensive Tissue Hypothesis: Independent Support from Highly Encephalized Fish." *Current Anthropology* 44:705–707.

Kay, R. F. 1975. "The Functional Adaptations of Primate Molar Teeth." *American Journal of Physical Anthropology* 42:195–215.

Kay, R. F., M. Cartmill, and M. Balow. 1998. "The Hypoglossal Canal and the Origin of Human Vocal Behaviour." *Proceedings of the National Academy of Sciences* 95:5417–5419.

Kelly, R. C. 1993. *Constructing Inequality: The Fabrication of a Hierarchy of Virtue Among the Etoro.* Ann Arbor: University of Michigan Press.

Kelly, R. L. 1995. *The Foraging Spectrum: Diversity in Hunter-Gatherer Lifeways.* Washington, DC: Smithsonian Institution.

Khaitovich, P., H. E. Lockstone, M. T. Wayland, T. M. Tsang, S. D. Jayatilaka, A. J. Guo, J. Zhou, M. Somel, L. W. Harris, E. Holmes, S. Pääbo, and S. Bahn. 2008. "Metabolic Changes in Schizophrenia and Human Brain Evolution." *Genome Biology* 9: R124, 1–11.

King, J. E. 2000. *Mayo Clinic on Digestive Health.* Rochester, MN: Mayo Clinic.

Klein, R. G. 1999. *The Human Career: Human Biological and Cultural Origins.* Chicago: University of Chicago Press.

Knott, C. 2001. "Female Reproductive Ecology of the Apes: Implications for Human Evolution." In *Reproductive Ecology and Human Evolution,* P. Ellison, ed., 429–463. New York: Aldine.

Koebnick, C., A. L. Garcia, P. C. Dagnelie, C. Strassner, J. Lindemans, N. Katz, C. Leitzmann, and I. Hoffmann. 2005. "Long-Term Consumption of a Raw Food Diet Is Associated with Favorable Serum LDL Cholesterol and Triglycerides but Also with Elevated Plasma Homocysteine and Low Serum HDL Cholesterol in Humans." *Journal of Nutrition* 135:2372–2378.

Koebnick, C., C. Strassner, I. Hoffmann, and C. Leitzmann. 1999. "Consequences of a Longterm Raw Food Diet on Body Weight and Menstruation: Results of a Questionnaire Survey." *Annals of Nutrition and Metabolism* 43:69–79.

Kuhn, S. L., and M. C. Stiner. 2006. "What's a Mother to Do? The Division of Labor Among Neandertals and Modern Humans in Eurasia." *Current Anthropology* 47:953–963.

Kummer, H. 1995. *In Quest of the Sacred Baboon: A Scientist's Journey.* Princeton, NJ: Princeton University Press.

Kuzawa, C. W. 1998. "Adipose Tissue in Human Infancy and Childhood: An Evolutionary Perspective." *Yearbook of Physical Anthropology* 41:177–209.

Laden, G., and R. W. Wrangham. 2005. "The Rise of the Hominids as an Adaptive Shift in Fallback Foods: Plant Underground Storage Organs (USOs) and Australopith Origins." *Journal of Human Evolution* 49:482–498.

Lancaster, J., and C. Lancaster. 1983. "Parental Investment, the Hominid Adaptation." In *How Humans Adapt: A Biocultural Odyssey,* D. S. Ortner, ed., 33–56. Washington, DC: Smithsonian Institution Press.

Langkilde, A. M., M. Champ, and H. Andersson. 2002. "Effects of High-Resistant-Starch Banana Flour (RS2) on In Vitro Fermentation and the Small-Bowel Excretion of Energy, Nutrients, and Sterols: An Ileostomy Study." *American Journal of Clinical Nutrition* 75:104–111.

Lawrie, R. A. 1991. *Meat Science,* 5th ed. Oxford, UK: Pergamon Press.

Leach, E. 1970. *Lévi-Strauss.* London: Fontana.

Lee, R. B., and I. DeVore. 1968. *Man the Hunter.* Cambridge, MA: Harvard University Press.

Lee, R. B. 1979. *The !Kung San: Men, Women and Work in a Foraging Society.* Cambridge, UK: Cambridge University Press.

Lee, S. W., J. H. Lee, S. H. Han, J. W. Lee, and C. Rhee. 2005. "Effect of Various Processing Methods on the Physical Properties of Cooked Rice and on *In Vitro* Starch Hydrolysis and Blood Glucose Response in Rats." *Starch-Starke* 57:531–539.

Leonard, W. R., and M. L. Robertson. 1997. "Comparative Primate Energetics and Hominid Evolution." *American Journal of Physical Anthropology* 102:265–281.

Leonard, W. R., J. J. Snodgrass, and M. L. Robertson. 2007. "Effects of Brain Evolution on Human Nutrition and Metabolism." *Annual Review of Nutrition* 27:311–327.

Lepowsky, M. 1993. *Fruit of the Motherland: Gender in an Egalitarian Society.* New York: Columbia University Press.

Letterman, J. B. 2003. *Survivors: True Tales of Endurance.* New York: Simon & Schuster.

Lévi-Strauss, C. 1969. *The Raw and the Cooked. Introduction to a Science of Mythology. I.* New York: Harper & Row.

Lewin, R., and R. A. Foley. 2004. *Principles of Human Evolution.* New York: Wiley-Blackwell.

Lieberman, D. E., G. E. Krovitz, F. W. Yates, M. Devlin, and M. St. Claire. 2004. "Effects of Food Processing on Masticatory Strain and Craniofacial Growth in a Retrognathic Face." *Journal of Human Evolution* 46:655–677.

Lieberman, D. E., B. M. McBratney, and G. Krovitz. 2002. "The Evolution and Development of Cranial Form in *Homo sapiens.*" *Proceedings of the National Academy of Sciences* 99:1134–1139.

Livesey, G. 1995. "The Impact of Complex Carbohydrates on Energy Balance." *European Journal of Clinical Nutrition* 49:S89–S96.

———. 2001. "A Perspective on Food Energy Standards for Nutrition Labelling." *British Journal of Nutrition* 85:271–287.

Low, B. 2000. *Why Sex Matters.* Princeton, NJ: Princeton University Press.

Lucas, P. 2004. *Dental Functional Morphology: How Teeth Work.* Cambridge, UK: Cambridge University Press.

Lucas, P. W., K. Y. Ang, Z. Sui, K. R. Agrawal, J. F. Prinz, and N. J. Dominy. 2006. "A Brief Review of the Recent Evolution of the Human Mouth in Physiological and Nutritional Contexts." *Physiology and Behavior* 89:36–38.

Mabjeesh, S. J., J. Galindez, O. Kroll, and A. Arieli. 2000. "The Effect of Roasting Nonlinted Whole Cottonseed on Milk Production by Dairy Cows." *Journal of Dairy Science* 83:2557–2563.

MacLarnon, A. M., R. D. Martin, D. J. Chivers, and C. M. Hladik. 1986. "Some Aspects of Gastro-Intestinal Allometry in Primates and Other Mammals." In *Definition et Origines de L'Homme*, M. Sakka, ed., 293–302. Paris: Editions du CNRS.

Mallol, C., F. W. Marlowe, B. M. Wood, and C. C. Porter. 2007. "Earth, Wind, and Fire: Ethnoarchaeological Signals of Hadza Fires." *Journal of Archaeological Science* 34:2035–2052.

Man, E. H. 1932 (1885). *On the Aboriginal Inhabitants of the Andaman Islands*. London: Royal Anthropological Institute of Great Britain and Ireland.

Mania, D. 1995. "The Earliest Occupation of Europe: The Elbe-Saale Region (Germany)." In *The Earliest Occupation of Europe*, W. Roebroeks and T. van Kolfschoten, eds., 85–102. Leiden, Netherlands: European Science Foundation.

Mania, D., and U. Mania. 2005. "The Natural and Socio-Cultural Environment of *Homo Erectus* at Bilzingsleben, Germany." In *The Hominid Individual in Context: Archaeological Investigations of Lower and Middle Palaeolithic Landscapes, Locales and Artefacts*, C. Gamble and M. Porr, eds., 98–114. London and New York: Routledge.

Marlowe, F. W. 2007. "Hunting and Gathering: The Human Sexual Division of Foraging Labor." *Cross-Cultural Research* 41:170–196.

———. 2003. "A Critical Period for Provisioning by Hadza Men: Implications for Pair Bonding." *Evolution and Human Behavior* 24:217–229.

Marshall, L. 1998 (1976). "Sharing, Talking, and Giving: Relief of Social Tensions Among the !Kung." In *Limited Wants, Unlimited Means: A Reader on Hunter-Gatherer Economics and the Environment*, J. M. Gowdy, ed., 65–85. Washington, DC: Island Press.

Marshall, W. E. 1873. *A Phrenologist Among the Todas, or the Study of a Primitive Tribe in South India: History, Character, Customs, Religion, Infanticide, Polyandry, Language*. London: Longmans, Green & Co.

Martin, R. D., D. J. Chivers, A. M. MacLarnon, and C. M. Hladik. 1985. "Gastrointestinal Allometry in Primates and Other Mammals." In *Size and Scaling in Primate Biology,* W. L. Jungers, ed., 61–89. New York: Plenum.

Mazza, P. P. A., F. Martini, B. Sala, M. Magi, M. P. Colombini, G. Giachi, F. Landucci, C. Lemorini, F. Modugno, and E. Ribechini. 2006. "A New Palaeolithic Discovery: Tar-Hafted Stone Tools in a European Mid-Pleistocene Bone-Bearing Bed." *Journal of Archaeological Science* 33:1310–1318.

McBrearty, S., and A. S. Brooks. 2000. "The Revolution That Wasn't: A New Interpretation of the Origin of Modern Human Behavior." *Journal of Human Evolution* 39:453–563.

McGee, H. 2004. *On Food and Cooking: The Science and Lore of the Kitchen.* New York: Scribners.

McHenry, H. M., and K. Coffing. 2000. "*Australopithecus* to *Homo*: Transformations in Body and Mind." *Annual Review of Anthropology* 29:125–146.

Medel, P., F. Baucells, M. I. Gracia, C. de Blas, and G. G. Mateos. 2002. "Processing of Barley and Enzyme Supplementation in Diets for Young Pigs." *Animal Feed Science and Technology* 95:113–122.

Medel, P., M. A. Latorre, C. de Blas, R. Lazaro, and G. G. Mateos. 2004. "Heat Processing of Cereals in Mash or Pellet Diets for Young Pigs." *Animal Feed Science and Technology* 113:127–140.

Megarry, T. 1995. *Society in Prehistory: The Origins of Human Culture.* New York: New York University Press.

Mehlman, P. T., and D. M. Doran. 2002. "Factors Influencing Western Gorilla Nest Construction at Mondika Research Center." *International Journal of Primatology* 23:1257–1285.

Melis, A. P., B. Hare, and M. Tomasello. 2006a. "Engineering Cooperation in Chimpanzees: Tolerance Constraints on Cooperation." *Animal Behavior* 72:275–286.

———. 2006b. "Chimpanzees Recruit the Best Collaborators." *Science* 311:1297–1300.

Merrill, A. L., and B. K. Watt. 1955. *Energy Value of Foods—Basis and Derivation. USDA Handbook No. 74.* Washington, DC: U.S. Department of Agriculture.

Meyer, J. H., J. Dressman, A. S. Fink, G. L. Amidon. 1985. "Effect of Size and Density on Canine Gastric Emptying of Nondigestible Solids." *Gastroenterology* 89:805–813.

Meyer, J. H., J. Elashoff, V. Porter-Fink, J. Dressman, and G. L. Amidon. 1988. "Human Postprandial Gastric Emptying of 1–3-millimeter Spheres." *Gastroenterology* 94:1315–1325.

Mill, J. S. 1966 (1869). "The Subjection of Women." In *Three Essays by J. S. Mill.* London: Oxford University Press.

Millett, K. 1970. *Sexual Politics.* New York: Doubleday.

Milton, K. 1987. "Primate Diets and Gut Morphology: Implications for Hominid Evolution." In *Food and Evolution: Towards a Theory of Human Food Habits,* M. Harris and E. B. Ross, eds., 93–115. Philadelphia: Temple University Press.

———. 1993. "Diet and Primate Evolution." *Scientific American* 269:86–93.

———. 1999. "A Hypothesis to Explain the Role of Meat-Eating in Human Evolution." *Evolutionary Anthropology* 8:11–21.

Milton, K., and M. W. Demment. 1988. "Chimpanzees Fed High and Low Fiber Diets and Comparison with Human Data." *Journal of Nutrition* 118:1082–1088.

Mitani, J. C., D. P. Watts, and M. N. Muller. 2002. "Recent Developments in the Study of Wild Chimpanzee Behavior." *Evolutionary Anthropology* 11:9–25.

Moggi-Cecchi, J. 2001. "Questions of Growth." *Nature* 414:596–597.

Mora, R., and I. de la Torre. 2005. "Percussion Tools in Olduvai Beds I and II (Tanzania): Implications for Early Human Activities." *Journal of Anthropological Archaeology* 24:179–192.

Muir, J. G., A. Birkett, I. Brown, G. Jones, and K. O'Dea. 1995. "Food Processing and Maize Variety Affects Amounts of Starch Escaping Digestion in the Small Intestine." *American Journal of Clinical Nutrition* 61:82–89.

Mulder, M. B., A. T. Kerr, and M. Moore. 1997. *Time Allocation Among the Kipsigis of Kenya.* New Haven, CT: Human Relations Area Files Inc.

Munroe, R. H., R. L. Munroe, J. A. Shwayder, and G. Arias. 1997. *Newar Time Allocation.* New Haven, CT: Human Relations Area Files Inc.

Munroe, R. L., and R. H. Munroe. 1990a. *Black Carib Time Allocation.* New Haven, CT: Human Relations Area Files Inc.

———. 1990b. *Samoan Time Allocation.* New Haven, CT: Human Relations Area Files Inc.
———. 1991. *Logoli Time Allocation.* New Haven, CT: Human Relations Area Files Inc.
Murakami, K., S. Sasaki, Y. Takahashi, K. Uenishi, M. Yamasaki, H. Hayabuchi, T. Goda, J. Oka, K. Baba, K. Ohki, T. Kohri, K. Muramatsu, and M. Furuki. 2007. "Hardness (Difficulty of Chewing) of the Habitual Diet in Relation to Body Mass Index and Waist Circumference in Free-Living Japanese Women Aged 18–22 y." *American Journal of Clinical Nutrition* 86:206–213.
Murdock, G. P., and C. Provost. 1973. "Factors in the Division of Labor by Sex: A Cross-Cultural Analysis. *Ethnology* 12:203–225.
Murgatroyd, S. 2002. *The Dig Tree.* London: Bloomsbury.
Nagalakshmi, D., V. R. B. Sastry, and D. K. Agrawal. 2003. "Relative Performance of Fattening Lambs on Raw and Processed Cottonseed Meal Incorporated Diets." *Asian-Australian Journal of Animal Science* 16:29–35.
Nishida, T., H. Ohigashi, and K. Koshimizu. 2000. "Tastes of Chimpanzee Plant Foods." *Current Anthropology* 41:431–465.
Noah, L., F. Guillon, B. Bouchet, A. Buleon, C. Molis, M. Gratas, and M. Champ. 1998. "Digestion of Carbohydrate from White Beans (*Phaseolus vulgaris* L.) in Healthy Humans." *Journal of Nutrition* 128:977–985.
Nunn, C. L., P. Lindenfors, E. R. Pursall, and J. Rolff. 2008. "On Sexual Dimorphism in Immune Function." *Philosophical Transactions of the Royal Society of London, Series B,* 364:61–69.
O'Connell, J. F., K. Hawkes, K. D. Lupo, and N. G. Blurton-Jones. 2002. "Male Strategies and Plio-Pleistocene Archaeology." *Journal of Human Evolution* 43:831–872.
O'Dea, K. 1991. "Traditional Diet and Food Preferences of Australian Aboriginal Hunter-Gatherers." *Philosophical Transactions of the Royal Society of London, Series B* 334:223–241.
Oakley, K. P. 1955. "Fire as a Paleolithic Tool and Weapon." *Proceedings of the Prehistoric Society* 21:36–48.
———. 1963. "On Man's Use of Fire, with Comments on Tool-Making and Hunting." In *Social Life of Early Man,* S. L. Washburn, ed., 176–193. London: Methuen.

———. 1962. "The Earliest Tool-Makers." In *Evolution und Hominisation*, G. Kurth, ed., 157–169. Stuttgart, Germany: Geburtstage von Gerehard Heberer.

Oka, K., A. Sakuarae, T. Fujise, H. Yoshimatsu, T. Sakata, and M. Nakata. 2003. "Food Texture Differences Affect Energy Metabolism in Rats." *Journal of Dental Research* 82:491–494.

Olkku, J., and C. Rha. 1978. "Gelatinisation of Starch and Wheat Flour Starch—A Review." *Food Chemistry* 3:293–317.

Onoda, H. 1974 (1999). *No Surrender: My Thirty Year War.* Annapolis, MD: U.S. Naval Institute Press.

Oosterwal, G. 1961. *People of the Tor: A Cultural-Anthropological Study on the Tribes of the Tor Territory (Northern Netherlands New-Guinea).* Assen, Netherlands: Van Gorcum.

Owen, J. B. 1991. *Cattle Feeding.* Ipswich, UK: Farming Press.

Pagel, M., and W. Bodmer. 2003. "A Naked Ape Would Have Fewer Parasites." *Proceedings of the Royal Society of London B (Suppl.)* 270:S117–S119.

Palmer, D. J., M. S. Gold, and M. Makrides. 2005. "Effect of Cooked and Raw Egg Consumption on Ovalbumin Content of Human Milk: A Randomized, Double-Blind, Cross-Over Trial." *Clinical and Experimental Allergy* 35:173–178.

Palmer, K. 2002. "Raw Food Best for Pets? Some Say Yes; Many Vets Say No." *Minneapolis Star Tribune*, August 5, 2002.

Pálsson, G. 2001. *Writing on Ice: the Ethnographic Notebooks of Vilhjalmur Stefansson.* Hanover, NH, and London: University Press of New England.

Panter-Brick, C. 2002. "Sexual Division of Labor: Energetic and Evolutionary Scenarios." *American Journal of Human Biology* 14:627–640.

Paolisso, M. J., and R. D. Sackett. 1988. *Time Allocation Among the Yukpa of Yurmutu.* New Haven, CT: Human Relations Area Files Inc.

Pastó, I., E. Allué, and J. Vasllverdú. 2000. "Mousterian Hearths at Abric Romaní, Catalonia (Spain)." In *Neanderthals on the Edge*, C. Stringer, R. Barton, and J. Finlayson, eds., 59–67. Oxford, UK: Oxbow Books.

Pate, D. 2006. "Hunter-Gatherer Social Complexity at Roonka Flat, South Australia." In *Social Archaeology of Indigenous Societies,* B. David, I. J. McNiven, and B. Barker, eds., 226–241. Canberra, Australia: Aboriginal Studies Press.

Pattanaik, A. K., V. R. B. Sastry, and R. C. Katiyar. 2000. "Effect of Thermal Processing of Cereal Grain on the Performance of Crossbred Calves Fed Starters Containing Protein Sources of Varying Ruminal Degradability." *Asian-Australian Journal of Animal Sciences* 13:1239–1244.

Perlès, C. 1979. "Les origines de la cuisine: L'acte alimentaire dans l'histoire de l'homme." *Communications* 31:4–14.

———. 1999. "Feeding Strategies in Prehistoric Times." In *Food: A Culinary History from Antiquity to the Present,* J.-L. Flandrin, and M. Montanari, eds., 21–31. New York: Columbia University Press.

Pettit, J. 1990. *Utes: the Mountain People.* Boulder, CO: Johnson Books.

Philbrick, N. 2000. *In the Heart of the Sea: The Tragedy of the Whaleship Essex.* New York: Viking.

Pleau, M. J., J. E. Huesing, G. P. Head, and D. J. Feir. 2002. "Development of an Artificial Diet for the Western Corn Rootworm." *Entomologia Experimentalis et Applicata* 105:1–11.

Plummer, T. 2004. "Flaked Stones and Old Bones: Biological and Cultural Evolution at the Dawn of Technology." *Yearbook of Physical Anthropology* 47: 118–164.

Pollan, M. 2008. *In Defense of Food: An Eater's Manifesto.* New York: Penguin.

Polo, M. 1926. *The Travels of Marco Polo (The Venetian).* New York: Boni & Liverwright.

Pond, C. M. 1998. *The Fats of Life.* Cambridge, UK: Cambridge University Press.

Pontzer, H., and R. W. Wrangham. 2004. "Climbing and the Daily Energy Cost of Locomotion in Wild Chimpanzees: Implications for Hominoid Locomotor Evolution." *Journal of Human Evolution* 46:315–333.

Potts, R. 1998. "Environmental Hypotheses of Hominin Evolution." *Yearbook of Physical Anthropology* 41:93–138.

Preece, R. C., J. A. J. Gowlett, S. A. Parfitt, D. R. Bridgland, and S. G. Lewis. 2006. "Humans in the Hoxnian: Habitat, Context and Fire Use at Beeches Pit, West Stow, Suffolk, UK." *Journal of Quaternary Science* 21:485–496.

Prince Peter, of Greece and Denmark. 1955. "The Todas: Some Additions and Corrections to W. H. R. Rivers's Book, Observed in the Field." *Man (N.S.)* 55:89–93.

Pruetz, J. D., and P. Bertolani. 2007. "Savanna Chimpanzees, *Pan troglodytes verus*, Hunt with Tools." *Current Biology* 17:1–6.

Pullen, A. G. 2005. "Fire and Cognition in the Paleolithic." Ph.D. diss., University of Cambridge.

Pusey, A. E., G. W. Oehlert, J. Williams, and J. Goodall. 2005. "Influence of Ecological and Social Factors on Body Mass of Wild Chimpanzees." *International Journal of Primatology* 26:3–31.

Radcliffe-Brown, A. 1922. *The Andaman Islanders: A Study in Social Anthropology.* Cambridge, UK: Cambridge University Press.

Raffaele, P. 2006. "Speaking Bonobo." *Smithsonian Magazine* 37:74.

Ragir, S. 2000. "Diet and Food Preparation: Rethinking Early Hominid Behavior." *Evolutionary Anthropology* 9:153–155.

Ragir, S., M. Rosenberg, and P. Tierno. 2000. "Gut Morphology and the Avoidance of Carrion Among Chimpanzees, Baboons, and Early Hominids." *Journal of Anthropological Research* 56:477–512.

Rao, M. A., and D. B. Lund. 1986. "Kinetics of Softening Foods: A Review." *Journal of Food Processing and Preservation* 10:311–329.

Read, P. P. 1974. *Alive: the Story of the Andes Survivors.* Philadelphia and New York: Lippincott.

Reznick, D. N., M. J. Bryant, D. Roff, C. K. Ghalambor, and D. E. Ghalambor. 2004. "Effect of Extrinsic Mortality on the Evolution of Senescence in Guppies." *Nature* 431:1095–1099.

Riches, D. 1987. "Violence, Peace and War in 'Early' Human Society: The Case of the Eskimo." In *The Sociology of War and Peace,* C. Creighton and M. Shaw, eds., 17–36. London: Macmillan.

Rightmire, G. P. 1998. "Human Evolution in the Mid Pleistocene: The Role of *Homo heidelbergensis*." *Evolutionary Anthropology* 6:218–227.

———. 2004. "Brain Size and Encephalization in Early to Mid-Pleistocene *Homo*." *American Journal of Physical Anthropology* 124:109–123.

Rivers, W. H. R. 1906. *The Todas.* London: Macmillan.
Roach, R. 2004. "Splendid Specimens: The History of Nutrition in Bodybuilding." *Wise Traditions* 5.
Robertson, D. 1973. *Survive the Savage Sea.* New York: Praeger.
Robinson, G. A. 1846. *Brief Report of an Expedition to the Aboriginal Tribes of the Interior . . . March to August 1846.* Melbourne, Australia: Manuscript in National Museum.
Rolff, J. 2002. "Bateman's Principle and Immunity." *Proceedings of the Royal Society of London, Series B* 269:867–872.
Rolls, E. T. 2005. "Taste, Olfactory, and Food Texture Processing in the Brain, and the Control of Food Intake." *Physiology and Behavior* 85:45–56.
Rombauer, I. S., and M. R. Becker. 1975. *Joy of Cooking.* New York: Bobbs-Merrill.
Rose, F. G. G. 1960. *Classification of Kin, Age Structure and Marriage Among the Groote Eylandt Aborigines: A Study in Method and a Theory of Australian Kinship.* Berlin: Akademie-Verlag.
Rosell, M., P. Appleby, and T. Key. 2005. "Height, Age at Menarche, Body Weight and Body Mass Index in Life-Long Vegetarians." *Public Health Nutrition* 8:870–875.
Rowlett, R. M. 1999. "'Comment' on Wrangham et al. (1999)." *Current Anthropology* 40:584–585.
Ruiz de Huidobro, F., E. Miguel, B. Blazquez, E. Onega. 2005. "A Comparison Between Two Methods (Warner–Bratzler and Texture Profile Analysis) for Testing Either Raw Meat or Cooked Meat." *Meat Science* 69:527–536.
Ruskin, J. 1902 (1865). *Sesame and Lilies.* New York: Homewood.
Rutherfurd, S. M., and P. J. Moughan. 1998. "The Digestible Amino Acid Composition of Several Milk Proteins: Application of a New Bioassay." *Journal of Dairy Science* 81:909–917.
Sannaveerappa, T., K. Ammu, and J. Joseph. 2004. "Protein-Related Changes During Salting of Milkfish (*Chanos chanos*)." *Journal of the Science of Food and Agriculture* 84:863–869.
Savage-Rumbaugh, S., and R. Lewin. 1994. *Kanzi: The Ape at the Brink of the Human Mind.* New York: Wiley.
Sawyer, G. J., V. Deak, E. Sarmiento, and R. Milner. 2007. *The Last Human: A Guide to Twenty-Two Species of Extinct Humans.* New Haven, CT: Yale University Press.

Schulze, L. G. 1891. "The Aborigines of the Upper and Middle Finke River: Their Habits and Customs, with Introductory Notes on the Physical and Natural-History Features of the Country." *Transactions and Proceedings and Reports of the Royal Society of South Australia* 14:210–246.

Secor, S. M. 2003. "Gastric Function and Its Contribution to the Postprandial Metabolic Response of the Burmese Python *Python molurus*." *Journal of Experimental Biology* 206:1621–1630.

———. 2009. "Specific Dynamic Action: A Review of the Postprandial Metabolic Response." *Journal of Comparative Physiology B*, in press.

Secor, S. M., and A. C. Faulkner. 2002. "Effects of Meal Size, Meal Type, Body Temperature, and Body Size on the Specific Dynamic Action of the Marine Toad, *Bufo marinus*." *Physiological and Biochemical Zoology* 75:557–571.

See, R., S. M. Abdullah, D. K. McGuire, A. Khera, M. J. Patel, J. B. Lindsey, S. M. Grundy, and J. A. De Lemos. 2007. "The Association of Differing Measures of Overweight and Obesity with Prevalent Atherosclerosis—The Dallas Heart Study." *Journal of the American College of Cardiology* 50:752–759.

Sergant, J., P. Crombé, and Y. Perdaen. 2006. "The 'Invisible' Hearths: A Contribution to the Discernment of Mesolithic Non-Structured Surface Hearths." *Journal of Archaeological Science* 33:999–1007.

Shelley, M. W. 1982 (1818). *Frankenstein or, The Modern Prometheus*. Chicago: University of Chicago Press.

Sherman, P. W., and J. Billing. 2006. "Darwinian Gastronomy: Why We Use Spices." *BioScience* 49:453–463.

Shultz, S., and R. I. M. Dunbar. 2007. "The Evolution of the Social Brain: Anthropoid Primates Contrast with Other Vertebrates." *Proceedings of the Royal Society of London, Series B* 274:2429–2436.

Silberbauer, G. B. 1981. *Hunter and Habitat in the Central Kalahari Desert*. Cambridge, UK: Cambridge University Press.

Sims, E. A., and E. J. Danforth. 1987. "Expenditure and Storage of Energy in Man." *Journal of Clinical Investigation* 79:1019–1025.

Sizer, F. S., and E. Whitney. 2006. *Nutrition: Concepts and Controversies*. Belmont, CA: Thomson/Wadsworth.

Smith, B. H. 1991. "Dental Development and the Evolution of Life History in Hominidae." *American Journal of Physical Anthropology* 86:157–174.

Smith, C. S., W. Martin, and K. A. Johansen. 2001. "Sego Lilies and Prehistoric Foragers: Return Rates, Pit Ovens, and Carbohydrates." *Journal of Archaeological Science* 28:169–183.

Smith, G. 1995. *Time Allocation Among the Madurese of Gedang-Gedang.* New Haven, CT: Human Relations Area Files Inc.

Smith, M. E., and D. G. Morton. 2001. *The Digestive System: Basic Science and Clinical Conditions.* London: Harcourt.

Smith, R. J., and W. L. Jungers. 1997. "Body Mass in Comparative Primatology." *Journal of Human Evolution* 32:523–559.

Southgate, D. A. T. 1981. *The Relationship Between Food Composition and Available Energy. Provisional Agenda Item 4.1.3, Joint FAO/WHO/UNU Expert Consultation on Energy and Protein Requirements, Rome, 5 to 17 October 1981.* Norwich, UK: A.R.C. Food Research Institute.

Southgate, D. A. T., and J. V. G. A. Durnin. 1970. "Calorie Conversion Factors—An Experimental Reassessment of the Factors Used in the Calculation of the Energy Value of Human Diets." *British Journal of Nutrition* 24:517–535.

Spencer, B. 1927. *The Arunta: a Study of a Stone Age People.* London: Macmillan.

Speth, J. D. 1989. "Early Hominid Hunting and Scavenging: The Role of Meat as an Energy Source." *Journal of Human Evolution* 18:329–343.

Sponheimer, M., B. H. Passey, D. J. de Ruiter, D. Guatelli-Steinberg, T. E. Cerling, and J. A. Lee-Thorp. 2006. "Isotopic Evidence for Dietary Variability in the Early Hominin *Paranthropus robustus*." *Science* 314:980–982.

Spoor, F., M. G. Leakey, P. N. Gathogo, F. H. Brown, S. C. Antón, I. McDougall, C. Kiarie, F. K. Manthi, and L. N. Leakey. 2007. "Implications of New Early *Homo* Fossils from Ileret, East of Lake Turkana, Kenya." *Nature* 448:688–691.

Stahl, A. B. 1989. "Comment on James (1989)." *Current Anthropology* 30:18–19.

Stanford, C. B. 1999. *The Hunting Apes: Meat Eating and the Origins of Human Behavior.* Princeton, NJ: Princeton University Press.

Stanford, C. B., and H. T. Bunn. 2001. *Meat-Eating and Human Evolution.* Oxford, UK: Oxford University Press.

Stead, S. M., and L. Laird. 2002. *Handbook of Salmon Farming.* London: Springer.

Stedman, H. H., B. W. Kozyak, A. Nelson, D. M. Thesier, L. T. Su, D. W. Low, C. R. Bridges, J. B. Shrager, N. Minugh-Purvis, and M. A. Mitchell. 2004. "Myosin Gene Mutation Correlates with Anatomical Changes in the Human Lineage." *Nature* 428:415–418.

Steele, J., and S. Shennan. 1996. "Darwinism and Collective Representations." In *The Archaeology of Human Ancestry: Power, Sex and Tradition,* J. Steele and S. Shennan, eds., 1–42. London: Routledge.

Stefansson, V. 1913. *My Life with the Eskimo.* New York: Macmillan.

———. 1944. *Arctic Manual.* New York: Macmillan.

Steward, J. H., and L. C. Faron. 1959. *Native Peoples of South America.* New York: McGraw-Hill.

Subias, S. M. 2002. "Cooking in Zooarchaeology: Is This Issue Still Raw?" In *Consuming Passions and Patterns of Consumption,* P. Miracle and N. Milner, eds., 7–16. Oxford, UK: Oxbow.

Svihus, B., A. K. Uhlen, and O. M. Harstad. 2005. "Effect of Starch Granule Structure, Associated Components and Processing on Nutritive Value of Cereal Starch: A Review." *Animal Feed Science and Technology* 122:303–320.

Symons, M. 1998. *A History of Cooks and Cooking.* Urbana and Chicago: University of Illinois Press.

Tanaka, J. 1980. *The San Hunter-Gatherers of the Kalahari: a Study in Ecological Anthropology.* Tokyo: University of Tokyo Press.

Tanaka, T., A. Mizumoto, N. Haga, and Z. Itoh. 1997. "A New Method to Measure Gastric Emptying in Conscious Dogs: A Validity Study and Effects of EM523 and L-NNA." *American Journal of Physiology-Gastrointestinal and Liver Physiology* 272:G909–G915.

Teaford, M. F., P. S. Ungar, and F. E. Grine. 2002. "Paleontological Evidence for the Diets of African Plio-Pleistocene Hominins with Special Reference to Early *Homo*." In *Human Diet: Its Origin and Evolution,* P. S. Ungar and M. F. Teaford, eds., 143–166. Westport, CT: Bergin & Garvey.

Tester, R. F., X. Qi, J. Karkalas. 2006. "Hydrolysis of Native Starches with Amylases." *Animal Feed Science and Technology* 130:39–54.

Thieme, H. 1997. "Lower Palaeolithic Hunting Spears from Germany." *Nature* 385:807–810.
———. 2000. "Lower Palaeolithic Hunting Weapons from Schöningen, Germany—The Oldest Spears in the World." *Acta Anthropologica Sinica* 19 (supplement): 140–147.
———. 2005. "The Lower Paleolithic Art of Hunting." In *The Hominid Individual in Context: Archaeological Investigations of Lower and Middle Paleolithic Landscapes, Locales and Artefacts*, C. S. Gamble and M. Parr, eds., 115–132. London: Routledge.
Thomas, E. M. 1959. *The Harmless People.* New York: Vintage Press.
Thompson, M. E., S. M. Kahlenberg, I. C. Gilby, and R. W. Wrangham. 2007. "Core Area Quality Is Associated with Variance in Reproductive Success Among Female Chimpanzees at Kanyawara, Kibale National Park." *Animal Behaviour* 73:501–512.
Tindale, N. B. 1974. *Aboriginal Tribes of Australia: Their Terrain, Environmental Controls, Distribution, Limits, and Proper Names. With an Appendix on Tasmanian Tribes by Rhys Jones.* Berkeley: University of California Press.
Tornberg, E. 1996. "Biological Aspects of Meat Toughness." *Meat Science* 43:S175–S191.
Toth, N., and K. Schick. 2006. *The Oldowan: Case Studies into the Earliest Stone Age.* Gosport, IN: Stone Age Institute Press.
Turnbull, C. 1962. *The Forest People.* New York: Simon & Schuster.
———. 1965. *Wayward Servants: The Two Worlds of the African Pygmies.* Westport, CT: Greenwood Press.
———. 1974 (1972). *The Mountain People.* London: Picador.
Tylor, E. B. 1870 (1964). *Researches into the Early History of Mankind.* Chicago: University of Chicago Press.
Ungar, P. 2004. "Dental Topography and Diets of *Australopithecus afarensis* and Early *Homo*." *Journal of Human Evolution* 46:605–622.
Ungar, P. S., F. E. Grine, and M. F. Teaford. 2006. "Diet in Early *Homo*: A Review of the Evidence and a New Model of Dietary Versatility." *Annual Review of Anthropology* 35:209–228.
U.S. Department of Agriculture, Agricultural Research Service. 2007. *USDA National Nutrient Database for Standard Reference, Release 21.* Nutrient Data Laboratory home page, www.ars.usda.gov/nutrientdata.

Valero, H., and E. Biocca. 1970. *Yanoáma: The Narrative of a White Girl Kidnapped by Amazonian Indians.* New York: E. P. Dutton.

Vlassara, H., W. Cai, J. Crandall, T. Goldberg, R. Oberstein, V. Dardaine, M. Peppa, and E. J. Rayfield. 2002. "Inflammatory Mediators Are Induced by Dietary Glycotoxins, a Major Risk Factor for Diabetic Angiopathy." *Proceedings of the National Academy of Sciences, USA* 99:15596–15601.

Wade, N. 2007. *Before the Dawn: Recovering the Lost History of Our Ancestors.* London: Penguin.

Waguespack, N. 2005. "The Organization of Male and Female Labor in Foraging Societies: Implications for Early Paleoindian Archaeology." *American Anthropologist* 107:666–676.

Waldron, K. W., M. L. Parker, and A. C. Smith. 2003. "Plant Cells Walls and Food Quality." *Comprehensive Reviews in Food Science and Food Safety* 2:101–119.

Walker, A., and P. Shipman. 1996. *The Wisdom of the Bones: In Search of Human Origins.* New York: Alfred A. Knopf.

Wandsnider, L. 1997. "The Roasted and the Boiled: Food Composition and Heat Treatment with Special Emphasis on Pit-Hearth Cooking." *Journal of Anthropological Archaeology* 16:1–48.

Ward, C. V. 2002. "Interpreting the Posture and Locomotion of *Australopithecus afarensis*: Where Do We Stand?" *Yearbook of Physical Anthropology* 45:185–215.

Washburn, S. L., and C. S. Lancaster. 1968. "The Evolution of Hunting." In *Man the Hunter,* R. B. Lee and I. DeVore, eds., 293–303. Cambridge, MA: Harvard University Press.

Watts, D. P., and J. C. Mitani. 2002. "Hunting Behavior of Chimpanzees at Ngogo, Kibale National Park, Uganda." *International Journal of Primatology* 23:1–28.

Weil, J. 1993. *Time Allocation Among Bolivian Quechua Coca Cultivators.* New Haven, CT: Human Relations Area Files Inc.

Weiner, J. 1994. *The Beak of the Finch: A Story of Evolution in Our Time.* New York: Knopf.

Wells, J. C. K. 2006. "The Evolution of Human Fatness and Susceptibility to Obesity: An Ethological Approach." *Biological Reviews* 81:183–205.

Werdelin, L., and M. E. Lewis. 2005. "Plio-Pleistocene Carnivora of Eastern Africa: Species Richness and Turnover Patterns." *Zoological Journal of the Linnean Society* 144:121–144.

Werner, D. 1993. *Mekranoti Time Allocation*. New Haven, CT: Human Relations Area Files Inc.

Westra, C. 2004. *How to Do the Raw Food Diet with Joy for Awesome Health and Success.* Published privately at www.IncreasedLife.com.

Wheeler, P. 1992. "The Influence of the Loss of Functional Body Hair on Hominid Energy and Water Budgets." *Journal of Human Evolution* 23:379–388.

White, T. D., B. Asfaw, D. DeGusta, H. Gilbert, G. D. Richards, G. Suwa, and F. C. Howell. 2003. "Pleistocene *Homo sapiens* from Middle Awash, Ethiopia." *Nature* 423:742–747.

Wiessner, P. 2002. "Hunting, Healing, and Hxaro Exchange: A Long-Term Perspective on !Kung (Ju/'hoansi) Large-Game Hunting." *Evolution and Human Behavior* 23:407–436.

Williams, J. M., A. E. Pusey, J. V. Carlis, B. P. Farm, and J. Goodall. 2002. "Female Competition and Male Territorial Behavior Influence Female Chimpanzees' Ranging Patterns." *Animal Behaviour* 63:347–360.

Wittig, R. M., and C. Boesch. 2003. "Food Competition and Linear Dominance Hierarchy Among Female Chimpanzees of the Tai National Park." *International Journal of Primatology* 24:847–867.

Wobber, V., B. Hare, and R. Wrangham. 2008. "Great Apes Prefer Cooked Food." *Journal of Human Evolution* 55:343–348.

Wolpoff, M. H. 1999. *Paleoanthropology,* 2nd ed. Boston: McGraw-Hill.

Wood, B., and D. Strait. 2004. "Patterns of Resource Use in Early *Homo* and *Paranthropus*." *Journal of Human Evolution* 46:119–162.

Wood, B. A., and M. Collard. 1999. "The Human Genus." *Science* 284:65–71.

Wood, W., and A. Eagly. 2002. "A Cross-Cultural Analysis of the Behavior of Women and Men: Implications for the Origins of Sex Differences." *Psychological Bulletin* 128:699–727.

Woodhead-Galloway, J. 1980. *Collagen: The Anatomy of a Protein.* London: Edwin Arnold.

Wrangham, R. 1977. "Feeding Behaviour of Chimpanzees in Gombe National Park, Tanzania." In *Primate Ecology,* T. H. Clutton-Brock, ed., 503–538. London: Academic Press.

———. 2006. "The Cooking Enigma." In *Early Hominin Diets: The Known, the Unknown, and the Unknowable,* P. Ungar, ed., 308–323. New York: Oxford University Press.

Wrangham, R. W., and N. L. Conklin-Brittain. 2003. "The Biological Significance of Cooking in Human Evolution." *Comparative Biochemistry and Physiology, Part A* 136:35–46.

Wrangham, R. W., J. H. Jones, G. Laden, D. Pilbeam, and N. L. Conklin-Brittain. 1999. "The Raw and the Stolen: Cooking and the Ecology of Human Origins." *Current Anthropology* 40:567–594.

Wrangham, R. W., and D. Pilbeam. 2001. "African Apes as Time Machines." In *All Apes Great and Small. Volume 1: Chimpanzees, Bonobos, and Gorillas,* B. M. F. Galdikas, N. Briggs, L. K. Sheeran, G. L. Shapiro, and J. Goodall, eds., 5–18. New York: Kluwer Academic/Plenum.

Wrangham, R. W., M. L. Wilson, and M. N. Muller. 2006. "Comparative Rates of Aggression in Chimpanzees and Humans." *Primates* 47:14–26.

Yanigasako, S. J. 1979. "Family and Household: The Analysis of Domestic Groups." *Annual Review of Anthropology* 8:161–205.

Yeakel, J. D., N. C. Bennett, P. L. Koch, and N. J. Dominy. 2007. "The Isotopic Ecology of African Mole Rats Informs Hypotheses on the Evolution of Human Diet." *Proceedings of the Royal Society of London B* 274:1723–1730.

Zimmer, C. 2005. *Smithsonian Intimate Guide to Human Origins.* New York: HarperCollins.

INDEX